北京重点保护
野生植物图谱

The Key Protected Wild Plants in Beijing

曹吉鑫　李进宇　谢　磊　王茂良　等　编著

Cao Jixin, Li Jinyu, Xie Lei, Wang Maoliang, et al.

中国林业出版社
China Forestry Publishing House

·北京·

BEIJING

图书在版编目（CIP）数据

北京重点保护野生植物图谱 / 曹吉鑫等编著.

北京：中国林业出版社，2025. 6. -- ISBN 978-7-5219-

3229-4

Ⅰ. Q948.521

中国国家版本馆CIP数据核字第2025GX7908号

责任编辑：刘香瑞

装帧设计：刘临川

出版发行：中国林业出版社

（100009，北京市西城区刘海胡同7号，电话010-83143545）

电子邮箱：36132881@qq.com

网址：https://www.cfph.net

印刷：北京雅昌艺术印刷有限公司

版次：2025年6月第1版

印次：2025年6月第1次

开本：880mm×1230mm　1/16

印张：7.5

字数：320千字

定价：98.00元

编著人员

曹吉鑫　李进宇　谢　磊　王茂良　苑　超　韩思雨　张钰舒
张晓川　刘　冰　马俊丽　赵安琪　夏　磊　罗悦心　李文贺
肖佳敏　李明阳　罗依可　吴环宇

摄影人员

谢　磊　刘　冰　李进宇　罗悦心　李文贺　肖佳敏　李明阳
罗依可　吴环宇　周　繇

Editors

Cao Jixin　Li Jinyu　Xie Lei　Wang Maoliang　Yuan Chao　Han Siyu
Zhang Yushu　Zhang Xiaochuan　Liu Bing　Ma Junli　Zhao Anqi　Xia Lei
Luo Yuexin　Li Wenhe　Xiao Jiamin　Li Mingyang　Luo Yike　Wu Huanyu

Photographers

Xie Lei　Liu Bing　Li Jinyu　Luo Yuexin　Li Wenhe　Xiao Jiamin
Li Mingyang　Luo Yike　Wu Huanyu　Zhou Yao

序

北京野生植物是北京重要的适生乡土植物资源，是北京生态建设的核心植物种质资源；这些植物资源不仅造就了北京生态建设的植物多样性，而且带来了与其共生的动物和微生物的多样性；这些生物的多样性又维护了北京多种生态系统的稳定与和谐。可见，北京野生植物的保护工作生态意义重大。此外，野生植物资源利用与效益发挥在北京生态建设中具有广阔利用空间和巨大潜力。

2021年9月，国家林业和草原局、农业农村部正式向社会公布了调整后的《国家重点保护野生植物名录》，为我国重点保护野生植物工作给予更加深入和细致的指导。在此背景下，北京市园林绿化科学研究院组织人员对北京及周边的野生植物资源进行了科学考察，系统梳理了北京野生植物资源，并进行了科学分类。为满足林业与园林生态建设工程对植物材料选用的需求，服务北京市生态保护与植物科普工作，团队进一步整理编撰了《北京重点保护野生植物图谱》。该图谱收录了北京市域内分布的15种国家重点保护野生植物和86种北京市重点保护野生植物，每种重点保护野生植物都配有精美的特征照片，同时还有形态描述、分布区域以及保护策略的中英文说明。该书图文并茂，可作为野生植物资源保护研究的参考资料，也可作为相关部门制定生态保护和建设规划的科学依据，还可作为高等院校相关专业的工具书和社会公众的科普图册。同时，作为中英文对照图书，该书的出版也有助于北京的国际交往中心功能建设，为国外友人进一步了解北京提供一个新的窗口。

我相信，本书的出版不仅能吸引更多的读者关注北京重点保护野生植物资源，更能激发大家踊跃投身到保护行动中，众人齐心协力，汇聚磅礴力量，共同推动北京生物多样性之都建设，携手促进人与自然和谐共生。

我衷心地祝贺作者，为这本书的出版而感到高兴，特此作序。

中国工程院院士

2025 年 2 月 20 日

Beijing's native wild plants are crucial indigenous resources, uniquely adapted to the local environment, and serve as essential genetic material for the city's ecological growth. These plant species contribute to the richness of floral biodiversity in Beijing's ecological projects and support a diverse range of associated fauna and microorganisms. Together, this biological diversity plays a key role in maintaining the stability and balance of such metropolitan ecosystems. As such, protecting wild plant species in Beijing is of significant ecological importance, while their strategic use and potential benefits offer broad opportunities within the city's ongoing ecological development plans.

On September 7, 2021, the National Forestry and Grassland Administration, along with the Ministry of Agriculture and Rural Affairs of the People's Republic of China, officially released the updated *List of National Key Protected Wild Plants*. This revised list offers more thorough and detailed guidance for the country's efforts to conserve key wild plant species. In response, the Beijing Academy of Forestry and Landscape Architecture conducted a comprehensive survey of wild plant resources in Beijing and its surrounding regions. This effort involved systematically cataloging the capital's wild flora and classifying it scientifically. To support the selection of plant materials for forestry and landscape ecological projects, and to aid in promoting plant science awareness among Beijing's residents, the academy compiled and published *The Key Protected Wild Plants in Beijing*. This publication features 15 nationally protected species and 86 species protected at the municipal level within Beijing. Each species is presented with high-resolution diagnostic images and detailed descriptions in both Chinese and English, covering their morphology, distribution, and conservation strategy suggestions. Richly illustrated and meticulously documented, this book serves as an authoritative reference for wild plant conservation research. It provides a scientific foundation for ecological protection planning and policy development by relevant authorities. Additionally, it acts as a resource for academic programs in related fields and as an accessible guide for public education on botany. As a bilingual (Chinese-English) reference, the publication enhances Beijing's role as an international exchange hub, offering global audiences a valuable opportunity to explore the city's ecological and cultural identity.

This publication is set to garner increased public awareness of Beijing's key protected wild plant resources, inspiring collective conservation efforts and encouraging widespread participation in advancing Beijing's goal of becoming a biodiversity capital. By bringing together various stakeholders, it will strengthen the drive for urban ecological balance and promote a shared vision for a sustainable future where humanity and nature thrive together.

I would like to offer my sincere congratulations to the author and celebrate the release of this exceptional work. It is a great honor for me to write this foreword.

YIN Weilun

Academician of the Chinese Academy of Engineering

February 20, 2025

野生植物是生态系统的重要组成部分，它们不仅为动物提供食物、栖息地和繁殖场所，还通过光合作用产生氧气、吸收二氧化碳，维持着地球的碳循环和生态平衡。此外，野生植物还具有重要的经济价值，如药用价值、食用价值和观赏价值等。然而，随着人类活动的日益频繁，野生植物的生存环境面临着前所未有的威胁。土地开发、污染排放、气候变化等因素，导致许多野生植物数量减少甚至灭绝。这不仅破坏了生态系统的稳定性，也威胁到了人类的生存环境。因此，保护野生植物多样性、维护生态平衡，已成为我们共同的责任和使命。

北京市作为我国的首都，拥有悠久的文化历史和丰富的自然资源。其地处华北平原北部，地形独特，山地众多，西部为太行山山脉，北部为燕山山脉，山区面积约占全市面积的62%。海拔超过或接近2000米的高山有百花山、东灵山、海坨山、雾灵山等。多样的自然地理格局和地形特征，孕育了丰富的野生植物资源，也使北京市成为世界上生物多样性最为丰富的首都城市之一。

随着生态文明建设工作的推进，我国生物多样性保护工作取得重要进展。2021年9月，经国务院批准，《国家重点保护野生植物名录》正式向社会发布。名录中，北京市分布的野生植物有15种。之后，为进一步加强首都生物多样性保护，北京市园林绿化局与北京市农业农村局发布了调整后的《北京市重点保护野生植物名录》，旨在进一步规范和强化对这些珍贵植物的保护工作。该名录中，共列出了44科67种和1类（所有野生兰科植物）重点保护的野生植物（不包括《国家重点保护野生植物名录》已涵盖的物种），覆盖了蕨类植物、裸子植物和被子植物等多个植物大类群。与之前的版本相比，新版《北京市重点保护野生植物名录》不再划分"北京市一级"和"北京市二级"保护级别，而是基于物种濒危程度和保护价值进行了综合评估和动态调整。新增了如睡菜、款冬、宝珠草等19种野生植物，同时对北五味子、胡桃楸、黄精等因保护成效显著而种群恢复稳定的物种进行了移除。该名录的发布对保护珍稀物种，维护首都生物多样性和独特性具有重要意义。

《北京重点保护野生植物图谱》正是为了向广大读者展示这些珍贵的生命形态，以及它们在自然与人类社会中的独特地位与价值。本书详细介绍了北京市内分布的国家重点保护野生植物和北京市级保护野生植物，通过图文并茂的形式，带领读者走进这些植物的神秘世界，感受它们的自然之美与生态价值。本书基于新颁布的《北京市重点保护野生植物名录》和《国家重点保护野生植物名录》中北京市分布物种进行编写，收录国家级和北京市级重点保护植物共计46科79属101种。

本书收录的国家级和北京市级重点保护植物各具特色，具有较高的科研与利用价值和多样的生态功能。①珍稀濒危的物种：在北京山区，分布有多种珍稀濒危野生植物，它们面临着环境压力和人类活动影响，存在一定的灭绝风险。例如，北京水毛茛是一种水生植物，分布于昌平、延庆、怀柔等地溪水中。由于生态环境变化，尤其受2023年洪水影响，这种植物的数量急剧减少，其生存受到了巨大的威胁。

②丰富多样的生态功能：北京的一些重点保护野生植物具有独特的生态功能，对维护生态平衡具有重要作用。例如，睡菜为北京市北部地区湿地生态系统的重要植物类群；草麻黄为旱区生态系统重要组成部分；华北落叶松为北京市中高海拔山地重要建群树种；槭叶铁线莲、房山紫堇为崖壁生态系统组成的重要类群等。③观赏与药用价值：北京的许多重点保护野生植物还具有观赏和药用价值。例如，兰科植物是一类美丽的花卉植物，它们的花形奇特，具有很高的观赏价值；刺五加、款冬、秦艽、红景天、金莲花等均为著名中药材。

本书的编写以科学性为原则，同时兼顾简洁性与科普功能。每个物种只列出接受名，未列其下异名，物种信息资料以《中国植物志》、《Flora of China》以及近年来的分类学研究成果为依据，并参考iplant等网络共享平台中的物种信息进行编排。植物科的界定与排列顺序以最新的系统发育分类结果为准。在本书编写过程中，北京师范大学郭延平教授给予了大力协助，中国科学院植物研究所的林秦文、刘冰老师提供了大量野外植物照片，在此表示感谢。

本书不仅是一本科学普及读物，更是一本倡导生态保护理念的行动指南，旨在激发读者对自然生态的好奇心和探索欲，提升公众对生物多样性保护的关注与参与度。在未来的日子里，让我们携手努力，共同为保护北京的野生植物和生态环境贡献自己的力量。

编著者

2024 年 1 月 20 日

Wild plants are an important part of the ecosystem, they not only provide food, habitat and breeding sites for animals, but also produce oxygen and absorb carbon dioxide through photosynthesis, maintaining the carbon cycle and ecological balance of the earth. In addition, wild plants have important economic values, such as medicinal, edible and ornamental. However, with the increasing human activities, the living environment of wild plants is facing unprecedented threats. Factors such as land development, pollution emissions, and climate change have all led to the decline or even extinction of many wild plant populations. This not only destabilizes the ecosystem, but also threatens the living environment of human beings. Therefore, it has become the global common responsibility and mission to protect wild plant diversity and to maintain ecological balance.

As the capital of China, Beijing has a long cultural history and rich natural resources. Located in the northern part of the North China Plain, Beijing has a unique topography with many mountains, including the Taihang Mountain Range in the west and the Yanshan Mountain Range in the north, with mountainous areas accounting for about 62% of the city's area. High mountains with an altitude of more than or close to 2,000 meters include Baihua Mountain, Dongling Mountain, Haituo Mountain, Wuling Mountain and so on. These natural geographic patterns and topographic features provide a variety of geographic environments for the diversity of plant species, nurturing a rich wild plant resources for Beijing and making the city one of the capital cities with the richest biodiversity in the world.

With the promotion of ecological civilization construction, China's biodiversity protection work has made important progress. In September 2021, approved by the State Council, *List of National Key Protected Wild Plants* was officially released to the public. The list includes 15 plant species distributed in Beijing. Subsequently, in order to strengthen the protection of the capital's biodiversity, the Beijing Municipal Forestry and Parks Bureau and the Beijing Municipal Bureau of Agriculture and Rural Development released the adjusted *Beijing Key Wild Plants List* in 2022, which aims to further standardize and strengthen the protection of these precious plants. Except the 15 species included in *List of National Key Protected Wild Plants*, this list enumerates additional 67 species of 44 families and one category of wild plants that are distributed in Beijing and should be under key protection; they cover a wide range of plant groups such as ferns, gymnosperms and angiosperms. Compared with the previous version, the current new *Beijing Key Wild Plants List* no longer divided the protection level into "Beijing Grade 1" and "Beijing Grade 2", but comprehensively assess and dynamically adjust the protection degrees of these plants based on their endangerment degree and conservation values. Nineteen species of wild plants were newly added, such as *Menyanthes trifoliate*, *Tussilago farfara* and *Disporum viridescens*, while some species were downgraded or removed from the new list due to their remarkable conservation results and stable population recovery, such as *Schisandra chinensis*, *Juglans mandshurica*, and *Polygonatum sibiricum*. The new list will be of great significance for saving these rare species and maintaining the diversity and uniqueness of species in the area of Beijing.

The present book *The Key Protected Wild Plants in Beijing* is intended to show readers these precious life forms, as well as their unique status and value in nature and human society. This book introduces in detail the national key protected wild plant species and Beijing municipally protected wild species distributed in Beijing, and lead readers to enter the mysterious world of these plants and feel their natural beauty and ecological value through the form of illustrations and text. This book is based on the newly promulgated *Beijing Key Wild Plants List* and List *of National Key Protected Wild Plants*, and includes 101 species of national and municipal protected plants of 46 families and 79 genera, which are described in details and nicely illustrated.

The state-level and Beijing city-level protected plants included in this book have their own characteristics, some of which have high science and utilization values and diverse ecological functions. (1) Rare and endangered species: In the mountainous areas of Beijing, there are a variety of rare and endangered wild plants which are facing more or less risk of extinction due to environmental pressure and the impact of human activities, and are at a certain risk of extinction. For example, *Ranunculus pekinensis* is an aquatic plant that is found in streams in Changping, Yanqing and Huairou. Due to changes in the ecological environment, especially the flood damage in 2023, the number of individuals of this plant has drastically decreased, making the species now under great threat. (2) Rich and diversified ecological functions: Some key protected wild plants in Beijing have unique ecological functions, which play an important role in maintaining ecological balance. For example, *Menyanthes trifoliate* is an important species in the wetland ecosystem in the northern part of Beijing; *Ephedra sinica* is an important component of the ecosystem in the dry zone; *Larix gmelinii* var. *principis-rupprechtii* is an important species in the middle and high altitude mountains of Beijing; *Clematis acerifolia* and *Corydalis fangshanensis* are important members in the ecosystem of the cliffs. (3) Ornamental and medicinal values: Many of the key protected wild plants in Beijing have ornamental and medicinal values. For example, Orchidaceae is a class of beautiful flowering plants with peculiar flower shapes and high ornamental value; *Eleutherococcus senticosus*, *Tussilago farfara*, *Gentiana macrophylla*, *Rhodiola rosea* and *Trollius chinensis* are all famous Chinese medicinal herbs.

For the compilation of this book, we strive for an integrated feature of science, conciseness and popularization. Each species is listed only by its accepted scientific name, but not by its synonyms, and the species information is organized based on *Flora Reipublicae Populaeris Sinicae* (the Chinese version of *Flora of China*), *Flora of China*, and the recent results in taxonomic researches; we also referred to iplant and other network sharing platforms. The definition and systematic order of plant families are based on the latest results of the relevant phylogenetic and systematic studies. We are grateful to Prof. Guo Yanping of Beijing Normal University for her great assistance in preparation of this book. We would also like to express our gratitude to Dr. Lin Qinwen and Dr. Liu Bing from the Institute of Botany, Chinese Academy of Sciences for providing a large number of photographs of plants in the field.

This book is not only a science popularization book, but also an action guide to advocate the concept of ecological conservation, aims to stimulate readers' curiosity and desire to explore the wild nature and its ecosystem, and to raise the society's concern and participation in the protection of biodiversity. In the days to come, let's join hands to contribute to the protection of wild plants and their ecological environments.

Authors

January 20, 2024

目录

❶ 北京市分布的国家重点保护野生植物

❷ 北京市重点保护野生植物

Contents

1 National Key Protected Wild Plants Distributed in Beijing

2 Wild Plants under Key Protection of Beijing Municipality

北京市分布的
国家重点保护野生植物

National Key Protected Wild Plants
Distributed in Beijing

001

中 文 名 **百花山葡萄**

学　　名 *Vitis baihuashanensis*

保护级别 国家一级保护植物

葡萄科
Vitaceae

葡萄属
Vitis

[形态描述] 落叶木质藤本，长 15～20m。小枝圆柱形，红褐色，被灰白色柔毛，其后脱落至几无毛。卷须先端二分叉。叶通常鸟足状复叶，具 5 小叶；小叶具细柄。中间小叶均为菱形，中部以下通常 3 深裂，边缘具粗齿；基部一对小叶斜卵形，稍小，2 深裂或全裂，上面光滑，下面沿叶脉疏被白色柔毛。圆锥花序。浆果球形，熟时紫黑色。

[北京分布] 门头沟区（东灵山，小龙门）；房山区（百花山）。

[保护策略] 维持其原生栖息地的完整性，避免在其栖息地内进行大规模的开发活动；针对百花山葡萄生长地制订保护措施，定期巡护；制订百花山葡萄的人工繁育计划。

Morphological Description: Deciduous woody vine, 15–20 meters long. Young stems cylindrical, reddish-brown, covered with grayish-white pubescence, which later sheds, leaving the stems nearly glabrous. Tendrils bifurcate at the tips. Leaves usually compound, palmate with 5 leaflets; leaflets with slender petiolules, the central leaflet rhomboid, often 3-lobed below the middle, with coarsely toothed margins; the basal pair of leaflets are oblique-ovate, smaller, 2-lobed or entire, with the upper surface smooth and the underside sparsely pubescent along the veins with white soft hairs. Inflorescence a panicle. Berries globose, turning purple-black when ripe.

Distribution in Beijing: Mentougou District: Dongling Mountain, Xiaolongmen; Fangshan District: Baihua Mountain.

Protection Strategies: Maintain the integrity of its native habitat by avoiding large-scale development activities within the habitat. Establish protective measures in the growth areas of *Vitis baihuashanensis* and conduct regular patrols. Design an artificial breeding program for *Vitis baihuashanensis*.

002

中文名 **轮叶贝母**
学　名 *Fritillaria maximowiczii*
保护级别　国家二级保护植物

百合科
Liliaceae

贝母属
Fritillaria

[形态描述]　植株高27～54厘米。鳞茎由4～5枚或更多鳞片组成，周围又有许多米粒状小鳞片，后者很容易脱落。叶条状或条状披针形，先端不卷曲，通常每3～6枚排成一轮，极少为二轮，向上有时还有1～2枚散生叶。花单朵，少有2朵，紫色，稍有黄色小方格；叶状苞片1枚，先端不卷；雄蕊长约为花被片的3/5；花药近基着，花丝无小乳突。蒴果，棱上有翅。花期6月。

[北京分布]　密云区（坡头，雾灵山）。

[保护策略]　轮叶贝母多见于高海拔地区，应避免对其栖息地的破坏性开发，以保持其原生态环境。为珍贵中药材，过度采挖已威胁到其自然种群，需在其分布区内加强巡查，采取有效的法律手段控制采挖行为。

Morphological description: The plant is 27–54 cm tall. The bulb is composed of 4–5 or more scales, surrounded by numerous grain-sized small scales that detach easily. Leaves linear or linear-lanceolate, not curled at apex, typically arranged in whorls of 3–6 (rarely in two whorls), occasionally with 1–2 scattered leaves above. Flowers solitary, rarely 2, purple with faint yellowish checkered patterns; bract leaf-like, 1, apex not curled; stamens about 3/5 as long as tepals; anthers nearly basifixed, filaments lacking papillae. Capsule winged along ridges. Fl. Jun.

Distribution in Beijing: Miyun District: Potou, Wuling Mountain.

Protection Strategies: *Fritillaria maximowiczii* is commonly found in high-altitude regions. It is important to prevent destructive development in its habitat and maintain its pristine environment. As a valuable traditional Chinese medicinal herb, overharvesting has threatened its natural populations. It is necessary to strengthen patrols within its distribution area and implement effective legal measures to control harvesting activities.

003	中文名 **紫点杓兰**	兰科 Orchidaceae
	学　名 *Cypripedium guttatum*	
	保护级别 国家二级保护植物	杓兰属 *Cypripedium*

[形态描述] 植株高15～25厘米，具细长而横走的根状茎。茎直立，被短柔毛和腺毛，基部具数枚鞘，顶端具叶。叶2枚，常对生；叶片椭圆形、卵形或卵状披针形，先端急尖或渐尖，背面脉上疏被短柔毛或近无毛，干后常变黑色或浅黑色。花序顶生，具1花；花序柄密被短柔毛和腺毛；花苞片叶状，卵状披针形，先端急尖或渐尖，边缘具细缘毛；花梗和子房被腺毛；花白色，具淡紫红色或淡褐红色斑；中萼片卵状椭圆形或宽卵状椭圆形，先端急尖或短渐尖，背面基部常疏被微柔毛；合萼片狭椭圆形，先端2浅裂；花瓣常近匙形；唇瓣深囊状，钵形或深碗状，多少近球形，具宽阔的囊口，囊口前方

几乎不具内折的边缘，囊底有毛。退化雄蕊卵状椭圆形，先端微凹或近截形，上面有细小的纵脊突，背面有较宽的龙骨状突起。蒴果近狭椭圆形，被微柔毛。花期5～7月，果期8～9月。

[北京分布] 门头沟区（百花山）；密云区（坡头）；怀柔区（喇叭沟门）。

[保护策略] 紫点杓兰通常生长在针叶林下、阔叶林边缘及沼泽附近等湿润地带，应限制其分布区域的开发活动。因其长期面临非法采集的威胁，应加大巡查力度，防止非法采挖行为。

Morphological description: The plant is 15–25 cm tall, with slender and creeping rhizomes. The stem is erect, covered with short pubescence and glandular hairs, bearing several sheaths at the base and leaves at the apex. Leaves 2, often opposite; blade elliptic, ovate, or ovate-lanceolate, acute or acuminate at apex, sparsely pubescent or nearly glabrous on the abaxial veins, turning black or grayish-black when dried. Inflorescences terminal, 1-flowered; peduncle densely pubescent and glandular hairy; floral bracts leaflike, ovate-lanceolate, apex acute or acuminate, margin finely ciliate; pedicel and ovary glandular hairy; flowers white, with pale purplish-red or pale maroon spots;middle sepal ovate-elliptic or broadly ovate-elliptic, apex acute or shortly acuminate, abaxially often sparsely puberulent at base; united sepals narrowly elliptic, apex 2-lobed; petals often subspatulate; Labellum deeply saccate, mantle-shaped or deeply bowl-shaped, subglobose, with broad saccule, margin scarcely inflexed anterior to saccule, base of saccule hairy. Staminodes ovate-elliptic, apex retuse or subtruncate, with minute longitudinal ridges above and broader keel-like projections abaxially. Capsule subnarrowly ellipsoid, puberulent. Fl. May–Jul, fr. Aug–Sep.

Distribution in Beijing: Mentougou District: Baihua Mountain; Miyun District: Potou; Huairou District: Labagoumen.

Protection Strategies: *Cypripedium guttatum* typically grows in moist areas such as coniferous forests, the edges of broadleaf forests, and near wetlands. It is essential to restrict development activities in these regions. *Cypripedium guttatum* has long faced the threat of illegal collection, so increased patrol efforts are necessary to prevent illegal harvesting.

004	中文名	**大花杓兰**	兰科 Orchidaceae
	学　名	*Cypripedium macranthos*	杓兰属
	保护级别	国家二级保护植物	*Cypripedium*

[形态描述] 植株高25～50厘米，具粗短的根状茎。茎直立，稍被短柔毛或变无毛，基部具数枚鞘，鞘上方具3～4枚叶。叶片椭圆形或椭圆状卵形，先端渐尖或近急尖，两面脉上略被短柔毛或变无毛，边缘有细缘毛。花序顶生，具1花，极罕2花；花苞片叶状，通常椭圆形，较少椭圆状披针形，先端短渐尖，两面脉上通常被微柔毛；花梗和子房无毛；花大，紫色、红色或粉红色，通常有暗色脉纹，极罕白色；中萼片宽卵状椭圆形或卵状椭圆形，先端渐尖，无毛；合萼片卵形，先端2浅裂；花瓣披针形，先端渐尖，不扭转，内表面基部具长柔毛；唇瓣深囊状，近球形或椭圆形；囊口较小，囊底有毛；退化雄

蕊卵状长圆形，基部无柄，背面无龙骨状突起。蒴果狭椭圆形，无毛。花期6～7月，果期8～9月。

[北京分布] 房山区（上方山）；门头沟区（百花山）；密云区（坡头）；怀柔区（喇叭沟门，孙栅子）。

[保护策略] 保护其自然栖息地，限制伐木、开垦等活动，维持原生植被结构；加强对其分布区的巡查和监控，防止人为采挖和破坏；大花杓兰依赖特定的菌根菌共生发芽和生长，应避免对土壤中微生物群落的破坏。

Morphological description: Plants 25–50 cm tall, with stout short rhizomes. Stem erect, slightly pubescent or glabrescent, with several sheaths at base and 3–4 leaves above sheaths. Leaf blade elliptic or elliptic-ovate, apex acuminate or subacute, both surfaces slightly pubescent or glabrescent on veins, margin finely ciliate. Inflorescences terminal, 1-flowered, rarely 2-flowered; floral bracts leaflike, usually elliptic, less often elliptic-lanceolate, apex shortly acuminate, usually puberulent on both surfaces of veins; pedicels and ovary glabrous; flowers large, purple, red, or pink, usually with dark veins, rarely white; middle sepal broadly ovate-elliptic or ovate-elliptic, apex acuminate, glabrous; united sepals ovate, apex bilobed; petals lanceolate, apex petals lanceolate, apex acuminate, not twisted, inner surface basally villous; labellum deeply saccate, subglobose or ellipsoid; mouth of sac smaller, base of sac hairy; staminodes ovate-oblong, base sessile, abaxially without keel. Capsule narrowly ellipsoid, glabrous. Fl. Jun–Jul, fr. Aug–Sep.

Distribution in Beijing: Fangshan District: Shangfang Mountain; Mentougou District: Baihua Mountain; Miyun District: Potou; Huairou District: Labagoumen, Sunzhazi.

Protection Strategies: Protect its natural habitat by restricting activities such as logging and land reclamation, and maintain the integrity of the native vegetation structure. Strengthen patrols and monitoring in its distribution areas to prevent human collection and destruction. *Cypripedium macranthos* relies on a specific mycorrhizal fungus symbiosis for germination and growth, so it is crucial to avoid disrupting the microbial community in the soil.

005

中 文 名 **山西杓兰**
学　　名 *Cypripedium shanxiense*
保护级别 国家二级保护植物

兰科
Orchidaceae

杓兰属
Cypripedium

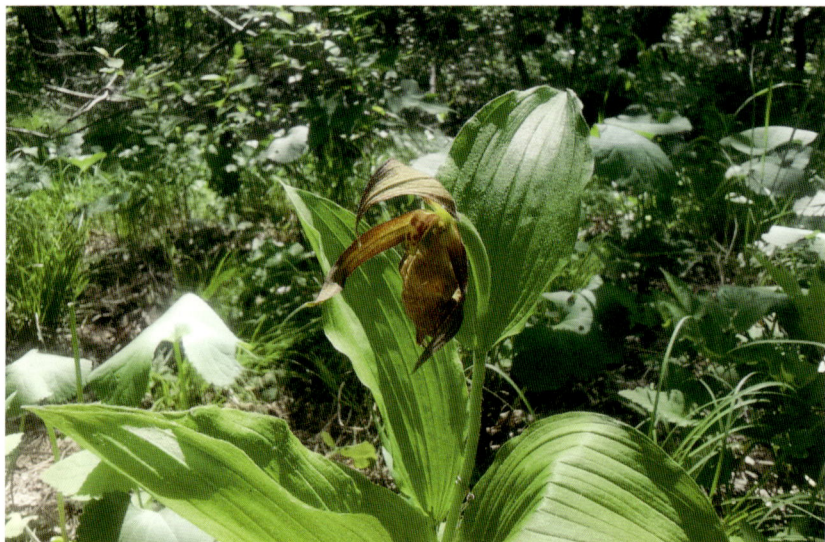

[形态描述] 植株高40～55厘米，具稍粗壮而匍匐的根状茎。茎直立，被短柔毛和腺毛，基部具数枚鞘，鞘上方具3～4枚叶。叶片椭圆形至卵状披针形，先端渐尖，两面脉上和背面基部有时有毛，边缘有缘毛。花序顶生，通常具2花，较少1花或3花；花序柄与花序轴被短柔毛和腺毛；花苞片叶状，两面脉上被疏柔毛；花梗和子房密被腺毛和短柔毛；花褐色至紫褐色，具深色脉纹，唇瓣常有深色斑点，退化雄蕊白色而有少数紫褐色斑点；中萼片披针形或卵状披针形，先端尾状渐尖，背面常有毛；合萼片与中萼片相似，先端深2裂，裂口深达5～10毫米；花瓣狭披针形或线形，先端渐尖，不扭转或稍扭转；唇瓣深囊状，近球形至椭圆形，囊底有毛，外面无毛；退化雄蕊长圆状椭圆形，基部有明显短柄。蒴果近梭形或狭椭圆形，疏被腺毛或变无毛。花期5～7月，果期7～8月。

[北京分布] 延庆区（玉渡山）。

[保护策略] 优先保护半阴湿润的林地和灌木丛等山西杓兰原生栖息地，严格限制开发活动，避免土地破坏；严格禁止在其栖息地进行采挖。

Morphological Description: Plants 40–55 cm tall, with slightly stout and creeping rhizomes. Stems erect, pubescent and glandular hairy, with several sheaths at base and 3–4 leaves above sheaths. Leaf blade elliptic to ovate-lanceolate, apex acuminate, both surfaces sometimes hairy on veins and abaxially at base, margin ciliate. Inflorescences terminal, usually 2-flowered, less often 1- or 3-flowered; peduncle and rachis pubescent and glandular hairy; floral bracts leaflike, sparsely pilose on both surfaces of veins; pedicel and ovary densely glandular hairy and pubescent; flowers brown to purplish-brown, with dark veins, labellum often darkly speckled, staminodes white and with a few purplish-brown spots; mid-salpiole lanceolate or ovate-lanceolate, apex caudate-acuminate, abaxially often hairy; sympodial sepal as long as middle sepal, often hairy. sepals similar to middle sepal, apex deeply 2-lobed, fissure 5–10 mm deep; petals narrowly lanceolate or linear, apex acuminate, not twisted or slightly twisted; labellum deeply saccate, subglobose to elliptic, base of sac hairy, outside glabrous; staminodes oblong-ellipsoid, base with conspicuous short stalk. Capsule subsessile or narrowly ellipsoid, sparsely glandular hairy or glabrescent. Fl. May–Jul, fr. Jul–Aug.

Distribution in Beijing: Yanqing District: Yudu Mountain.

Protection Strategies: Prioritize the protection of semi-shaded, humid woodlands and shrubland as their natural habitats, strictly limit development activities, and avoid land degradation. Prohibit any excavation activities within its habitat.

006

中文名 **手参**
学　名 *Gymnadenia conopsea*
保护级别 国家二级保护植物

兰科
Orchidaceae

手参属
Gymnadenia

[形态描述] 植株高20～60厘米。块茎椭圆形，肉质，下部掌状分裂，裂片细长。茎直立，圆柱形，基部具2～3枚筒状鞘，其上具4～5枚叶，上部具1至数枚苞片状小叶。叶片线状披针形、狭长圆形或带形，先端渐尖或稍钝，基部收狭成抱茎的鞘。总状花序具多数密生的花，圆柱形；花苞片披针形，直立伸展，先端长渐尖成尾状；子房纺锤形，顶部稍弧曲；花粉红色，罕为粉白色；中萼片宽椭圆形或宽卵状椭圆形，先端急尖，略呈兜状，具3脉；侧萼片斜卵形，反折，边缘向外卷，较中萼片稍长或几等长，先端急尖，具3脉，前面的1条脉常具支脉；花瓣直立，斜卵状三角形，边缘具细锯齿，先端

急尖，具3脉，前面的1条脉常具支脉，与中萼片相靠；唇瓣向前伸展，宽倒卵形，前部3裂，中裂片较侧裂片大，三角形，先端钝或急尖；距细而长，狭圆筒形，下垂，稍向前弯，向末端略增粗或略渐狭，长于子房；花粉团卵球形，具细长的柄和粘盘，粘盘线状披针形。花期6~8月。

[北京分布] 门头沟区；怀柔区；平谷区；密云区。

[保护策略] 保护其生长环境的原生植被和水源，减少基础设施建设带来的影响；加强手参栖息地的巡查，严格打击非法采挖和交易行为。

Morphological Description: Plants 20–60 cm tall. Tuber ellipsoid, fleshy, proximally palmately divided, lobes slender. Stem erect, terete, basally with 2–3 tubular sheaths with 4–5 leaves, distally with 1 to several bractlike leaflets. Leaf blade linear-lanceolate, narrowly oblong or strap-shaped, apex acuminate or slightly obtuse, base narrowed into an amplexicaul sheath. Racemes with numerous dense flowers, terete; floral bracts lanceolate, erect and spreading, apex long acuminate to caudate; ovary fusiform, slightly arcuate apically; flowers pink, rarely pink-white; middle sepal broadly elliptic or broadly ovate-elliptic, apex acute, slightly cucullate, 3-veined; lateral sepals obliquely ovate, reflexed, margins rolled outward, slightly longer than or several times equal to middle sepal, apex acute, 3-veined, anterior 1 vein often with a branch vein; flowers often with a branch vein; lateral sepals oblique ovate, reflexed, margins rolled outward, slightly or almost equal to middle sepal, apex acute, 3-veined, anterior 1 vein often with a branch vein. often tertiary; petals erect, obliquely ovate-triangular, margin serrulate, apex acute, 3-veined, anterior 1 vein often tertiary, abutting middle sepal; labellum spreading forward, broadly obovate, anteriorly 3-lobed, middle lobe larger than lateral lobes, triangular, apex obtuse or acute; spur thin and long, narrowly cylindrical, pendulous, slightly curved forward, slightly thickened or slightly attenuate toward end, longer than ovary; pollen mass ovoid Pollen cones ovoid, with slender stipe and mucro, mucro linear-lanceolate. Fl. Jun–Aug.

Distribution in Beijing: Mentougou District; Huairou District; Pinggu District; Miyun District.

Protection Strategies: Protect native vegetation and water sources in its growing environment, reduce the impact of infrastructure development. Strengthen patrols in the habitat of *Gymnadenia conopsea* and strictly combat illegal collection and trade activities.

007

中 文 名 **北京水毛茛**

学　　名 *Batrachium pekinense*

保护级别　国家二级保护植物

毛茛科
Ranunculaceae

毛茛属
Ranunculus

[**形态描述**]　多年生沉水草本。茎长30厘米以上，无毛或在节上有疏毛，分枝。叶有柄；叶片轮廓楔形或宽楔形，二型，沉水叶裂片丝形，上部浮水叶2～3回3～5中裂至深裂，裂片较宽，末回裂片短线形，无毛；叶柄基部有鞘，无毛或在鞘上有疏短柔毛。花梗无毛；萼片近椭圆形，有白色膜质边缘，脱落；花瓣白色，宽倒卵形，基部有短爪，蜜槽呈点状；雄蕊约15；花托有毛。花期5～8月。

[**北京分布**]　延庆区；怀柔区；昌平区（南口至居庸关一带）。

[**保护策略**]　减少人为干扰；在其原生栖息地采取生态修复措施；建立水质监测系统；以人工方式提高其种群数量，培育出的植株可在适宜季节移植到原生栖息地。

Morphological Description: Perennial submerged herb. Stem more than 30 cm long, glabrous or sparsely hairy at nodes, branched. Leaves petiolate; leaf blade cuneate or broadly cuneate in outline, dimorphic, submerged leaf segments filiform, upper floating leaves 2–3-ternately 3–5-cleft to deeply cleft, segments broader, ultimate segments shortly linear, glabrous; petiole base sheathed, glabrous or sparsely pubescent on sheath. Pedicel glabrous; sepals subelliptic,, with white membranous margin, deciduous; petals white, broadly obovate, base shortly clawed, nectaries punctate; stamens ca. 15; receptacle hairy. Fl. May-Aug.

Distribution in Beijing: Yanqing District; Huairou District; Changping District: the area from Nankou to Juyongguan.

Protection Strategies: Reduce human disturbance. Implement ecological restoration measures in its native habitat. Set up a water quality monitoring system. Increase its population artificially, and the cultivated plants can be transplanted to the native habitat during the appropriate season.

008

中文名 **槭叶铁线莲**
学　名 *Clematis acerifolia*
保护级别 国家二级保护植物

毛茛科
Ranunculaceae

铁线莲属
Clematis

[形态描述] 直立小灌木，高30~60厘米，除心皮外其余无毛。根木质，粗壮。老枝外皮灰色，有环状裂痕。叶为单叶，与花簇生；叶片五角形，基部浅心形，通常为不等的掌状5浅裂，中裂片近卵形，侧裂片近三角形，边缘疏生缺刻状粗牙齿。花2~4朵簇生；萼片5~8，开展，白色或带粉红色，狭倒卵形至椭圆形，无毛；雄蕊无毛；子房有柔毛。花期4月，果期5~6月。

[北京分布] 房山区（上方山，蒲洼，大安山，霞云岭）；门头沟区（丁家滩，太子墓，马套村）。

[保护策略] 严格控制生境中的采石和采矿等开发活动，减少对其栖息地的直接影响；土壤侵蚀会影响其生长环境，应在其分布区域采取防风固土措施，减少水土流失；收集种子并进行人工繁育，以增加其种群数量。

Morphological Description: Erect small shrubs, 30–60 cm tall, glabrous except for carpels. Root woody, stout. Older branches with grey outer bark, annularly fissured. Leaves simple, clustered with flowers; leaf blade pentagonal, base shallowly cordate, usually unequally palmately 5-lobed, middle lobe subovate, lateral lobes subtriangular, margin sparsely notched coarsely dentate. Flowers in clusters of 2–4; sepals 5–8, spreading, white or pinkish, narrowly obovate to elliptic, glabrous; stamens glabrous; ovary pilose. Fl. Apr, fr. May–Jun.

Distribution in Beijing: Fangshan District: Shangfang Mountain, Puwa, Da'an Mountain, Xiayunling; Mentougou District: Dingjiatan, Taizimu, Matao Village.

Protection Strategies: Strictly control development activities such as quarrying and mining in its habitat to minimize direct impacts on its environment. Soil erosion can affect its growing conditions; implement windbreak and soil conservation measures in its distribution area to reduce soil and water loss. Collect seeds and carry out artificial propagation to increase their population size.

009

中 文 名　**红景天**

学　　名　*Rhodiola rosea*

保护级别　国家二级保护植物

景天科
Crassulaceae

红景天属
Rhodiola

[形态描述]　多年生草本。根粗壮，直立。根颈短，先端被鳞片。花茎高20～30厘米。叶疏生，长圆形至椭圆状倒披针形或长圆状宽卵形，先端急尖或渐尖，全缘或上部有少数牙齿，基部稍抱茎。花序伞房状，密集多花；雌雄异株；萼片4，披针状线形，钝；花瓣4，黄绿色，线状倒披针形或长圆形，钝；雄花中雄蕊8，较花瓣长；鳞片4，长圆形，上部稍狭，先端有齿状微缺；雌花中心皮4，花柱外弯。蓇葖披针形或线状披针形，直立；种子披针形，一侧有狭翅。花期4～6月，果期7～9月。

[北京分布]　密云区；门头沟区（百花山，东灵山）。

[保护策略]　在主要分布区实施禁采措施，增加执法力度；在红景天生长的高海拔地区，应减少道路和旅游基础设施建设，避免栖息地的进一步破坏；在高山地区采取适当的土壤和植被保护措施。

Morphological Description: Herbs perennial. Roots stout, erect. Root neck short, apex scaly. Flowering stems 20–30 cm tall. Leaves sparsely, oblong to elliptic-oblanceolate or oblong-broadly ovate, apex acute or acuminate, entire or with few teeth distally, base slightly clasping. Inflorescences corymbose, densely many flowered; dioecious; sepals 4, lanceolate-linear, obtuse; petals 4, yellowish-green, linear-oblanceolate or oblong, obtuse; stamens 8 in male flowers, longer than petals; scales 4, oblong, slightly narrowed distally, apex toothed emarginate; female flowers with 4 central cortexes, styles externally curved. Follicles lanceolate or linear-lanceolate, erect; seeds lanceolate, narrowly winged on one side. Fl. Apr–Jun, fr. Jul–Sep.

Distribution in Beijing: Miyun District; Mentougou District: Baihua Mountain, Dongling Mountain.

Protection Strategies: Implement harvesting bans in key distribution areas and strengthen law enforcement. In high-altitude areas where *Rhodiola rosea* grows, reduce the construction of roads and tourism infrastructure to prevent further habitat destruction. Implement appropriate soil and vegetation protection measures in mountainous regions.

010

中文名 **甘草**
学　名 *Glycyrrhiza uralensis*
保护级别 国家二级保护植物

豆科
Fabaceae

甘草属
Glycyrrhiza

[形态描述] 多年生草本；根与根状茎粗壮，外皮褐色，里面淡黄色。茎直立，多分枝，密被鳞片状腺点、刺毛状腺体及白色或褐色的绒毛；托叶三角状披针形，两面密被白色短柔毛；小叶5～17枚，卵形、长卵形或近圆形，两面均密被黄褐色腺点及短柔毛，顶端钝，具短尖，基部圆，边缘全缘或微呈波状。总状花序腋生，具多数花，总花梗短于叶，密生褐色的鳞片状腺点和短柔毛；苞片长圆状披针形，褐色；花萼钟状，基部偏斜并膨大呈囊状，萼齿5，上部2齿大部分连合；花冠紫色、白色或黄色，旗瓣长圆形，顶端

微凹，基部具短瓣柄，翼瓣短于旗瓣，龙骨瓣短于翼瓣；子房密被刺毛状腺体。荚果弯曲呈镰刀状或呈环状，密集成球，密生瘤状突起和刺毛状腺体。种子3～11，暗绿色，圆形或肾形。花期6～8月，果期7～10月。

[北京分布] 昌平区（沙河镇）；门头沟区（西山）。

[保护策略] 针对甘草野生资源分布区，应实施禁采政策。

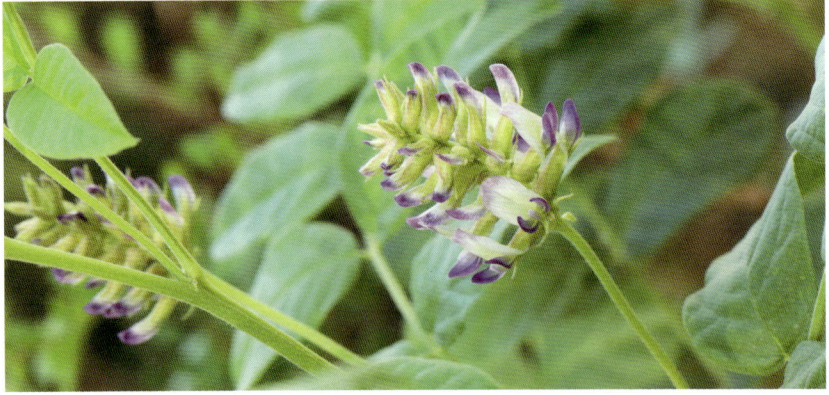

Morphological Description: Herbs perennial; roots and rhizomes coarsely rhizomatous, outer bark brown, inside yellowish. Stem erect, much branched, densely covered with scalelike glandular dots, prickly hairy glands, and white or brownish tomentum; stipules triangular-lanceolate, both surfaces densely white pubescent; leaflets 5–17, ovate, long ovate, or suborbicular, both surfaces densely yellow-brown glandular dots and pubescent, apex obtuse, mucronate, base rounded, margin entire or slightly undulate. Racemes axillary, with numerous flowers, the common pedicel shorter than the leaves, densely covered with brown scalelike glandular dots and pubescence; bracts oblong-lanceolate, brown; calyx campanulate, base oblique and inflated saccate, calyx teeth 5, the upper 2 teeth mostly conjoined; corolla purple, white, or yellow, the flagellum oblong, apical part concave, the base with a short petiole, the pterygoid petals shorter than the flagellum, the keel petals shorter than the pterygoid petals; ovary densely covered with prickly-hairlike glands. Pods curved sickle-shaped or ring-shaped, densely globose, densely tuberculate and prickly hairy glands. Seeds 3–11, dark green, orbicular or reniform. Fl. Jun–Aug, fr. Jul–Oct.

Distribution in Beijing: Changping District: Shahe Town; Mentougou District: Xishan.

Protection Strategies: Implement harvesting bans in areas where wild *Glycyrrhiza uralensis* resources are distributed.

011

中 文 名　**软枣猕猴桃**
学　　名　*Actinidia arguta*
保护级别　国家二级保护植物

狝猴桃科
Actinidiaceae

狝猴桃属
Actinidia

[形态描述] 大型落叶藤本；小枝基本无毛，皮孔长圆形至短条形；髓白色至淡褐色。叶卵形、长圆形、阔卵形至近圆形，顶端急短尖，基部圆形至浅心形，边缘具繁密的锐锯齿，横脉和网状小脉细，不发达，侧脉稀疏，6～7对。花序腋生或腋外生，为1～2回分枝，1～7花，或厚或薄地被淡褐色短绒毛，苞片线形。花绿白色或黄绿色；萼片4～6枚，卵圆形至长圆形，边缘较薄，两面薄被粉末状短茸毛；花瓣4～6片，楔状倒卵形或瓢状倒阔卵形，1花4瓣的其中有1片二裂至半；花丝丝状，花药黑色或暗紫色，长圆形箭头状；子房瓶状。果圆球形至柱状长圆形，有喙或喙不显著，无毛，无斑点，不具宿存萼片，成熟时绿黄色或紫红色。

[北京分布] 门头沟区（百花山，色树坟村，小龙门）；怀柔区（喇叭沟门，铁矿峪，渤海镇三岔村）；密云区（坡头，雾灵山，花园村，云蒙山，遥桥峪）；昌平区（大岭沟，南口镇，虎峪，双龙山）；房山区（十渡，蒲洼）；平谷区（千佛崖）；延庆区（千家店，松山）。

[保护策略] 设立野外采挖禁令，加强巡查，打击非法采挖行为；在保护区内应维持原有植被，提供适合其生长的湿润土壤和荫蔽环境；在适合的环境下推广软枣猕猴桃的人工种植。

Morphological Description: Large deciduous vine; branchlets essentially glabrous, lenticels oblong to short-striate; pith white to pale brown. Leaves ovate, oblong, broadly ovate to suborbicular, apically acute mucronate, base rounded to shallowly cordate, margin densely sharply serrate, transverse veins and reticulate veinlets fine, undeveloped, lateral veins sparse, 6–7 pairs. Inflorescences axillary or extra-axillary, 1–2-branched, 1–7-flowered, either thickly or thinly covered with pale brown downy hairs, bracts linear. Flowers greenish white or yellowish green; sepals 4–6; ovate-orbicular to oblong, margin thin, both surfaces thinly powdery velutinous; petals 4–6, cuneate-obovate or sarcocarpous-obroad-ovate, 1 of 4 in 1 flower 2-cleft to halfway; filaments filamentous, anthers black or dark purple, oblong-sagittate; ovary urceolate. Fruit globose to columnar-oblong, beaked or beak inconspicuous, glabrous and spotless, without persistent sepals, and greenish yellow or purplish red at maturity.

Distribution in Beijing: Mentougou District: Baihua Mountain, Seshufen Village, Xiaolongmen; Huairou District: Labagoumen, Tiekuangyu, Sancha Village in Bohai Town; Miyun District: Potou, Wuling Mountain, Huayuan Village, Yunmeng Mountain, Yaoqiaoyu; Changping District: Dalinggou, Nankou Town, Huyu, Shuanglong Mountain; Fangshan District: Shidu, Puwa; Pinggu District: Qianfo Cliff (Thousand Buddha Cliff);

Yanqing District: Qianjiadian, Songshan Mountain.

Protection Strategies: Implement a ban on wild harvesting and strengthen patrols to combat illegal digging. In protected areas, maintain the original vegetation and provide moist soil and shaded conditions suitable for growth. Promote the artificial cultivation of *Actinidia arguta* in suitable environments.

012	中文名 **丁香叶忍冬**	忍冬科 Caprifoliaceae
	学　名 *Lonicera oblata*	忍冬属
	保护级别 国家二级保护植物	*Lonicera*

[形态描述] 落叶灌木，高达2米；幼枝浅褐色，略呈四角形，老枝灰褐色；凡幼枝、芽的外鳞片、叶上面中脉和叶下面、叶柄、总花梗及苞片外面均被疏或密的短腺毛。冬芽有2对卵形、顶长尖的外鳞片。叶三角状宽卵形至菱状宽卵形，顶端短凸尖而钝头或钝形，基部宽楔形至截形。总花梗出自当年小枝的叶腋；苞片钻形，长达萼筒之半或不到；相邻两萼筒分离，无毛，萼檐杯状，齿不明显。果实红色，圆形；种子近圆形或卵圆形，稍扁平，淡棕褐色。果熟期7月。

[北京分布] 延庆区（松山）；门头沟区（龙门涧）。

[保护策略] 避免栖息地被开发利用，从而保持其自然生态环境；对已被破坏的生境进行生态恢复，改善生长环境，提升生物多样性；定期监测丁香叶忍冬的种群数量和分布；人工繁育丁香叶忍冬，有效增加其种群数量。

Morphological Description: Deciduous shrubs up to 2 m tall; young branches light brown, slightly quadrangular, old branches grey-brown; all young branches, outer scales of buds, midvein above and below leaves, petioles, pedicels and bracts sparsely or densely covered with short glandular hairs outside. Winter buds with 2 pairs of ovate, apically long pointed outer scales. Leaves triangular-broadly ovate to rhombic-broadly ovate, apically shortly convex and blunt or obtuse, base broadly cuneate to truncate. Pedicel from leaf axils of current year's branchlets; bracts subulate, up to half or less of calyx tube; two adjacent calyx tubes separate, glabrous, calyx limb cup-shaped, teeth inconspicuous. Fruit red, orbicular; seeds suborbicular or ovoid, slightly flattened, pale tan. Fr. Jul.

Distribution in Beijing: Yanqing District: Songshan Mountain; Mentougou District: Longmenjian Canyon.

Protection Strategies: Avoid the development and utilization of habitats to preserve the natural ecological environment. Implement ecological restoration for degraded habitats to improve the growing environment and enhance biodiversity. Regularly monitor the population size and distribution of *Lonicera oblata*. Artificial breeding of *Lonicera oblata* can effectively increase its population size.

013

中文名 **野大豆**
学　名 *Glycine soja*
保护级别 国家二级保护植物

豆科
Fabaceae

大豆属
Glycine

[形态描述] 一年生缠绕草本，长1~4米。茎、小枝纤细，全体疏被褐色长硬毛。叶具3小叶；托叶卵状披针形，急尖，被黄色柔毛。顶生小叶卵圆形或卵状披针形，先端锐尖至钝圆，基部近圆形，全缘，两面均被绢状糙伏毛，侧生小叶斜卵状披针形。总状花序通常短；花小；花梗密生黄色长硬毛；苞片披针形；花萼钟状，密生长毛，裂片5，三角状披针形，先端锐尖；花冠淡红紫色或白色，旗瓣近圆形，先端微凹，基部具短瓣柄，翼瓣斜倒卵形，有明显的耳，龙骨瓣比旗瓣及翼瓣短小，密被长毛；花柱短而向一侧弯曲。荚果长圆形，稍弯，两侧稍扁，密被长硬毛，种子间稍缢缩，干时易裂；种子2~3颗，椭圆形，稍扁，褐色至黑色。花期7~8月，果期8~10月。

[北京分布] 密云区；延庆区；怀柔区；昌平区；平谷区；房山区；门头沟

区；海淀区；顺义区；通州区；石景山区；丰台区。

[保护策略] 建立野大豆种质资源库，收集和保存不同地区的种子样本，以确保其基因多样性。

Morphological Description: Annual twining herb, 1–4 m long. Stem and branchlets slender, all sparsely brown hirsute. Leaves 3-foliolate; stipules ovate-lanceolate, acute, yellow pilose. Terminal leaflet ovate-orbicular or ovate-lanceolate, apex acute to obtuse-rounded, base subrounded, entire, both surfaces silky strigose, lateral leaflets obliquely ovate-lanceolate. Racemes usually short; flowers small; pedicels densely yellow hirsute; bracts lanceolate; calyx campanulate, densely hirsute, lobes 5, triangular-lanceolate, apex acute; corolla light reddish purple or white, flag petal suborbicular, apex retuse, base with short petiole, winged petals obliquely obovate, with conspicuous auricles, keel petals shorter than flag petals and winged petals, densely hirsute; styles short and curved to one side. Pods oblong, slightly curved, slightly flattened on both sides, densely hirsute, slightly constricted between seeds, easily cracked when dry; 2–3 seeds, ellipsoid, slightly flattened, brown to black. Fl. Jul–Aug, fr. Aug–Oct.

Distribution in Beijing: Miyun District; Yanqing District; Huairou District; Changping District; Pinggu District; Fangshan District; Mengtougou District; Haidian District; Shunyi District; Tongzhou District; Shijingshan District; Fengtai District.

Protection Strategies: Establish a germplasm resource bank for *Glycine soja* collecting and preserving seed samples from different regions to ensure genetic diversity.

014

中文名 **黄檗**

学　名 *Phellodendron amurense*

保护级别 国家二级保护植物

芸香科
Rutaceae

黄檗属
Phellodendron

[形态描述] 落叶乔木，树高10～20米。枝扩展，成年树的树皮有厚木栓层，浅灰或灰褐色，深沟状或不规则网状开裂，内皮薄，鲜黄色，味苦，黏质，小枝暗紫红色，无毛。叶轴及叶柄均纤细，有小叶5～13片，小叶薄纸质或纸质，卵状披针形或卵形，顶部长渐尖，基部阔楔形，一侧斜尖，或为圆形，叶缘有细钝齿和缘毛，叶面无毛或中脉有疏短毛，叶背仅基部中脉两侧密被长柔毛。花序顶生；萼片细小，阔卵形；花瓣紫绿色；雄花的雄蕊比花瓣长，退化雌蕊短小。果圆球形，蓝黑色；种子通常5粒。花期5～6月，果期9～10月。

[北京分布] 密云区（下营）；平谷区；延庆区（海坨山）；怀柔区（喇叭沟门）；昌平区（沟崖）；门头沟区（百花山）；房山区。

[保护策略] 在黄檗自然分布区限制人类活动，保护其生长环境；对破坏的栖息地进行生态恢复；建立黄檗可持续采集管理体系；收集和保存不同来源的黄檗种子及其繁殖材料，维护其遗传多样性。

Morphological Description: Deciduous tree, 10–20 m tall. Branches extended, the bark of adult trees has a thick corky layer, light grey or grey-brown, deeply furrowed or irregularly webbed open, the inner bark is thin, bright yellow, bitter, sticky, branchlets dark purplish red, glabrous. Leaf axis and petiole are slender, with 5–13 leaflets, leaflets thinly papery or papery, ovate-lanceolate or ovate, apex long acuminate, base broadly cuneate, one side obliquely pointed, or rounded, leaf margin with fine obtuse teeth and ciliate hairs, leaf surface is glabrous or with sparse short hairs in midvein, leaf abaxial surface is only densely villous on both sides of the midvein of the base. Inflorescences terminal; sepals minute, broadly ovate; petals purplish-green; stamens of male flowers longer than petals, pistillode shorter. Fruit globose, blue-black; seeds usually 5. Fl. May–

Jun, fr. Sep–Oct.

Distribution in Beijing: Miyun District: Xiaying; Pinggu District; Yanqing District: Haituo Mountain; Huairou District: Labagoumen; Changping District: Gou'ai Forest Farm; Mentougou District: Baihua Mountain; Fangshan District.

Protection Strategies: Limit human activities in the natural distribution zones of Phellodendron amurense, and protect its growing environment. Implement ecological restoration for damaged habitats. Establish a sustainable harvesting management system for *Phellodendron amurense*. Collect and preserve seeds and propagation materials from different sources to maintain its genetic diversity.

015

中文名 **紫椴**
学　名 *Tilia amurensis*
保护级别 国家二级保护植物

锦葵科
Malvaceae

椴属
Tilia

[形态描述] 落叶乔木，高25米，树皮暗灰色，片状脱落；嫩枝初时有白丝毛，很快变秃净，顶芽无毛，有鳞苞3片。叶阔卵形或卵圆形，先端急尖或渐尖，基部心形，上面无毛，下面浅绿色，脉腋内有毛丛，侧脉4～5对，边缘有锯齿，叶柄纤细，无毛。聚伞花序纤细，无毛，有花3～20朵；苞片狭带形，两面均无毛，下半部或下部1/3与花序柄合生；萼片阔披针形，外面有星状柔毛；退化雄蕊不存在；雄蕊较少，约20枚；子房有毛。果实卵圆形，被星状茸毛，有棱或棱不明显。花期7月。

[北京分布] 房山区（大安山）；门头沟区（妙峰山，东灵山，黑豆沟，小龙门，东松树峪，百花山）；昌平区（百合村）；怀柔区（喇叭沟门）；海淀区（香山）。

[保护策略] 保护其生长环境，维护生态系统的稳定性；对受到破坏的栖息地进行生态恢复；收集和保存不同来源的紫椴种子及繁殖材料，以维护其遗传多样性。

Morphological Description: Deciduous tree, 25 m tall, bark dark grey, flaky; shoots with white silky hairs at first, soon becoming bald and clean, terminal buds glabrous, with 3 scaly buds. Leaves broadly ovate or ovate-orbicular, apex acute or acuminate, base cordate, glabrous above, light green below, with tufts of hairs in the axils of the veins, 4–5 pairs of lateral veins, margins serrate; petiole slender, glabrous. Cymes slender, glabrous, with 3–20 flowers; bracts narrowly strap-shaped, glabrous on both surfaces, the lower half or lower 1/3 united with the inflorescence stalk; sepals broadly lanceolate, outside stellate pilose; staminodes absent; stamens fewer, ca. 20; ovary hairy. Fruit ovoid, stellate velutinous, with conspicuous or inconspicuous ridges. Fl. Jul.

Distribution in Beijing: Fangshan District: Da'an Mountain; Mentougou District: Miaofeng Mountain, Donglingshan, Heidou Valley, Xiaolongmen, Dong Songshuyu, Baihua Mountain; Changping District: Baihe Village; Huairou District: Labagoumen; Haidian District: Xiangshan.

Protection Strategies: Protect its growing environment and maintain the stability of the ecosystem. Implement ecological restoration for damaged habitats. Collect and preserve seeds from different sources to maintain its genetic diversity.

北京市重点保护
野生植物

Wild Plants under Key Protection of
Beijing Municipality

001

中文名 **小阴地蕨**
学　名 *Botrychium lunaria*

瓶尔小草科
Ophioglossaceae

小阴地蕨属
Botrychium

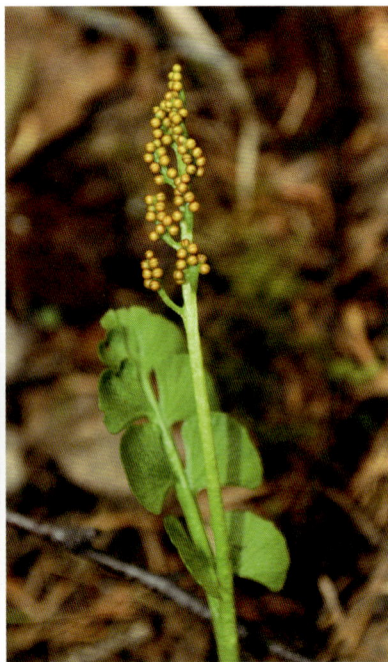

[形态描述] 根状茎短而直立。总叶柄基部有棕色托叶状苞片，宿存。不育叶片为阔披针形，圆头或圆钝头，一回羽状，羽片4～6对，对生或近于对生，彼此密接，扇形、肾圆形或半圆形，基部楔形，无柄，与中轴多少合生，外边缘全缘，或波状或多少分裂。叶脉呈扇形分离。孢子叶自不育叶片的基部抽出，孢子囊穗2～3次分裂，为狭圆锥形。直立，光滑无毛；孢子囊无柄且大。

[北京分布] 密云区（坡头，雾灵山）；怀柔区（喇叭沟门）；门头沟区（百花山）。

[保护策略] 定期开展野外调查，了解小阴地蕨种群的现状及变化趋势；通过研究小阴地蕨对气候变化的敏感性，确保其能在未来气候条件下生存；通过植被恢复、补植本地植物等手段，恢复其需要的阴湿环境；防治外来入侵植物；防治病虫害。

Morphological Description: Rhizome short and erect. Petiole base with brown stipule-like bracts, persistent. Sterile fronds broadly lanceolate with rounded or obtuse apex, once pinnate; pinnae 4–6 pairs, opposite or subopposite, closely spaced, flabellate, reniform, or semicircular, cuneate base, sessile, slightly adnate to rachis; outer margins entire, undulate, or slightly lobed. Venation fan-shaped and divergent. Fertile fronds arising from the base of sterile fronds; sporangia-bearing spikes 2–3 times divided, narrowly conical. Erect, glabrous; sporangia sessile and large.

Distribution in Beijing: Miyun District: Potou, Wuling Mountain; Huairou District: Labagoumen; Mentougou District: Baihua Mountain.

Protection Strategies: Conduct regular field surveys to monitor the status and trends of *Botrychium lunaria* populations. Study its climate sensitivity to ensure future survival. Restore its required shady, humid habitat through vegetation recovery and native plant reintroduction. Prevent invasive alien plants; control plant pests and diseases.

002

中文名 **北美球子蕨**

学　名 *Onoclea sensibilis*

球子蕨科
Onocleaceae

球子蕨属
Onoclea

[形态描述] 植株高30～70厘米。根状茎长而横走，暗褐色，疏被鳞片；鳞片棕色，阔卵形，长约5毫米，膜质，边缘全缘或微波状，先端渐尖。不育叶柄长20～50厘米，除基部栗褐色外呈禾秆色，疏被棕色鳞片；不育叶片基部羽裂，顶部羽裂，干后暗绿色，卵状三角形或卵形，纸质，微被鳞片，成熟时光滑无毛；羽片5～8对，狭椭圆形，基部1～2对最大，具短柄，边缘波状或浅裂，向上无柄或贴生，由叶轴翅相连；叶脉网状。能育叶柄长20～45厘米；能育叶片二回羽状，强烈收缩；羽片狭线形；小羽片球形，坚硬，彼此分离。囊群盖膜质，后缘基部着生，前端游离。

[北京分布] 延庆区（大庄科乡）；怀柔区（孙栅子，喇叭沟门）；门头沟区（百花山）。

[保护策略] 维持土壤阴湿、较低直射光和适宜空气湿度；避免人为干扰、垃圾堆积等直接影响；建立固定监测点，长期记录北美球子蕨的种群数量、气候条件、土壤湿度等信息；研究其对气候变化的适应能力，以指导后续的管理和栖息地改造。

Morphological Description: The plant grows 30–70cm tall. Rhizome long creeping, dark brown, sparsely scaly; scales brown, broadly ovate, ca. 5 mm, membranous, margin entire or slightly undulate, apex acuminate. Stipe of sterile frond 20–50 cm, stramineous except at chestnut-brown base, sparsely brown scaly; sterile lamina basally pinnatifid, apically pinnatifid, dark green when dry, ovate-triangular or ovate, papery, slightly scaly, glabrous when mature; pinnae 5–8 pairs, narrowly elliptic, basal 1 or 2 pairs largest,shortly stalked, undulate or lobed at margin, sessile or adnate upward, connected by rachis wing; veins reticulate. Stipe of fertile frond 20–45 cm; fertile lamina bipinnate, much contracted; pinnae narrowly linear; pinnules globose, hardened, separate from each other. Indusium membranous, fixed at posterior base, free distally.

Distribution in Beijing: Yanqing District: Dazhuangke Township; Huairou District: Sunzhazi, Labagoumen; Mentougou District: Baihua Mountain.

Protection Strategies: Maintain shady and humid soil conditions, low direct sunlight, and appropriate air humidity. Avoid direct human interference, such as litter accumulation. Establish fixed monitoring points to record long-term data on *Onoclea sensibilis* populations, climate conditions, and soil moisture. Study its adaptability to climate change to guide future management and habitat restoration efforts.

003

中文名 **小叶中国蕨**

学　名 *Aleuritopteris albofusca*

凤尾蕨科
Pteridaceae

粉背蕨属
Aleuritopteris

[形态描述] 植株高7～16厘米。根状茎短而直立，被栗黑色而有棕色狭边的披针形鳞片。叶簇生；柄长4～12厘米，栗黑色或栗红色，有光泽，基部疏被狭卵状披针形鳞片；叶片五角形，三裂，中央羽片最大，近菱形，二回羽状深裂；小羽片4～5对，斜展，基部一对最大，线状披针形，先端钝或急尖；裂片6～9对，长圆形或三角形，钝头或钝尖头，全缘。叶干后革质，上面暗绿色，平滑无毛；下面被腺体，分泌白色蜡质粉末。孢子囊群生小脉顶端，囊群盖膜质，淡棕色至褐棕色，连续，通常较阔，边缘具不整齐的浅波状圆齿。

[北京分布] 门头沟区（龙门涧）；怀柔区；密云区（坡头）；房山区（上方山）。

[保护策略] 小叶中国蕨生长在阴湿的环境中，应在其栖息地增加水源保护措施；在合适的人工生态环境中引种，选择适宜的湿润、遮阴环境，以建立更为分散的小种群，提高整体存活率；控制外来入侵植物。

Morphological Description: The plant is 7–16 cm tall. Rhizome short, erect, densely covered with lanceolate scales chestnut-black with narrow brown margins. Fronds clustered; stipe 4–12 cm, chestnut-black or chestnut-red, lustrous, sparsely covered with narrowly ovate-lanceolate scales at base. Lamina pentagonal, tripartite, central pinna largest, subrhombic, bipinnatifid to deeply bipinnatisect; pinnules 4 or 5 pairs, spreading obliquely, basal pair largest, linear-lanceolate, apex obtuse or acute; segments 6–9 pairs, oblong or triangular, apex obtuse or obtusely acute, entire. Lamina leathery when dry, adaxially dark green, smooth, glabrous; abaxially glandular, secreting white waxy powder. Sori terminal on veinlets; indusium membranous, pale brown to brownish-brown, continuous, usually broad, margin irregularly crenate-undulate.

Distribution in Beijing: Mentougou District: Longmenjian Canyon; Huairou District; Miyun District: Potou; Fangshan District: Shangfang Mountain.

Protection Strategies: Enhance water conservation measures in the shady and humid habitats where *Aleuritopteris albofusca* grows. Introduce the species into suitable artificial ecological environments with moist and shaded conditions to establish more dispersed small populations, improving overall survival rates. Control invasive alien plants.

004

中文名 **木贼麻黄**

学　名 *Ephedra equisetina*

麻黄科
Ephedraceae

麻黄属
Ephedra

[形态描述] 直立小灌木，高达1米，木质茎粗长，直立，稀部分匍匐状；小枝细，节间短，纵槽纹细浅（不明显），常被白粉，呈蓝绿色或灰绿色。叶2裂，褐色。雄球花单生或3～4个集生于节上，卵圆形或窄卵圆形，苞片3～4对，雄蕊6～8；雌球花常2个对生于节上，苞片3对，菱形或卵状菱形。雌球花成熟时肉质红色，长卵圆形或卵圆形，具短梗；种子通常1粒，窄长卵圆形。花期6～7月，种子8～9月成熟。

[北京分布] 石景山区（八宝山）；门头沟区（东灵山，小龙门）；延庆区（张山营，松山，兰角沟，塘子沟，八达岭，海坨山）。

[保护策略] 加强北京干旱、半干旱地区的环境保护，防止土地荒漠化和风蚀；减少人为活动对其栖息地的干扰；通过营造自然植被屏障，保持土壤的原始状态；防止过度采摘；推动人工种植。

Morphological Description: Erect small shrubs, up to 1 m tall; woody stems thick and long, erect, rarely partially procumbent; branchlets thin, internodes short, longitudinal grooves fine and shallow (indistinct), often pruinose, bluish green or grayish green. Leaves 2-lobed, brown. Male cones solitary or 3–4 clustered at nodes, ovoid or narrowly ovoid, bracts 3–4 pairs, stamens 6–8; female cones often 2 opposite at nodes, bracts 3 pairs, rhombic or ovate-rhombic. Female cones at maturity fleshy red, long-ovoid or ovoid, with short stalk; seeds usually 1, narrowly long-ovoid. Pollination Jun–Jul, seeds ripening Aug–Sep.

Distribution in Beijing: Shijingshan District: Babaoshan; Mentougou District: Dongling Mountain, Xiaolongmen; Yanqing District: Zhangshanying, Lanjiaogou, Songshan Mountain, Tangzigou, Badaling, Haituo Mountain.

Protection Strategies: Strengthen environmental protection in Beijing's arid and semi-arid regions to prevent desertification and wind erosion. Minimize human interference with habitats. Maintain the soil's original state by creating natural vegetation barriers. Prevent overharvesting and promote artificial cultivation.

005

中文名 **草麻黄**

学　名 *Ephedra sinica*

麻黄科
Ephedraceae

麻黄属
Ephedra

[形态描述] 草本状灌木，高20~40厘米；木质茎短或呈匍匐状，小枝直伸或微曲，表面细纵槽纹常不明显。叶2裂。雄球花多呈复穗状，常具总梗，苞片通常4对；雌球花单生，在幼枝上顶生，在老枝上腋生，卵圆形或矩圆状卵圆形，苞片4对。雌球花成熟时肉质红色；种子通常2粒，包于苞片内，黑红色或灰褐色。花期5~6月，种子8~9月成熟。

[北京分布] 延庆区；昌平区；门头沟区。

[保护策略] 避免北京干旱或半干旱区域的土地开发或过度利用；采取保护措施保持土壤表层结构，防止水土流失；规范药用草麻黄的采集行为；深入研究草麻黄的耐旱机制及其防风固沙作用，优化其栖息地的管理方式。

Morphological Description: Herbaceous shrubs, 20–40cm tall; woody stems short or prostrate; branchlets straight or slightly curved, shallowly furrowed. Leaves opposite. Pollen cones sessile or pedunculate, solitary or in clusters at nodes, rarely terminal; bracts in 4 pairs. Seed cones terminal or axillary, solitary, oblong-ovoid or subglobose; bracts in 4 pairs, red and fleshy at maturity. Seeds usually 2, enclosed in bracts, dark red or grayish brown. Pollination May–Jun, seed maturity Aug–Sep.

Distribution in Beijing: Yanqing District; Changping District; Mentougou District.

Protection Strategies: Avoid land development or overutilization in Beijing's arid and semi-arid regions. Implement protective measures to maintain the soil's surface structure and prevent erosion. Regulate the collection of the medicinal plant *Ephedra sinica*. Conduct in-depth research on its drought resistance mechanisms and its role in sand fixation and wind prevention, optimizing habitat management strategies.

006

中文名 **单子麻黄**

学　名 *Ephedra monosperma*

麻黄科
Ephedraceae

麻黄属
Ephedra

[形态描述] 草本状矮小灌木，高5～15厘米；木质茎短小，多分枝，弯曲并有节结状突起；绿色小枝开展或稍开展，常微弯曲，节间细短。叶2片对生，裂片短三角形。雄球花生于小枝上下各部，单生枝顶或对生节上，苞片3～4对，假花被较苞片长，雄蕊7～8；雌球花单生或对生节上，无梗，苞片3对。雌球花成熟时肉质红色，微被白粉，卵圆形或矩圆状卵圆形；种子外露，多为1粒。花期6月，种子8月成熟。

[北京分布] 门头沟区（东灵山）；延庆区（张山营）。

[保护策略] 采取封禁措施，避免土地开垦、工程建设等活动对其栖息地的破坏；限制野外采摘行为，推广人工种植单子麻黄；在干旱地区进行引种恢复。

Morphological Description: Herbaceous dwarf shrubs, 5–15 cm tall; woody stems short, much branched, nodes knotted; green branchlets spreading, usually slightly curved, slender, internodes short. Leaves opposite, free part shortly triangular. Pollen cones 2 or 3 and sessile or subsessile at nodes, rarely solitary at apex of branchlets; bracts in 3 or 4 pairs; exserted, with 7–8 sessile anthers. Seed cones solitary or opposite at nodes, sessile, ovoid to oblong-ovoid at maturity; bracts in 3 pairs, red and fleshy at maturity. Seeds exposed, mostly 1. Pollination Jun, seed maturity Aug.

Distribution in Beijing: Mentougou District: Dongling Mountain; Yanqing District: Zhangshanying.

Protection Strategies: Implement habitat protection measures to prevent damage caused by land reclamation, construction, and other activities. Restrict wild harvesting and promote the artificial cultivation of *Ephedra monosperma*. Conduct introduction and restoration efforts in arid regions.

007 中文名 **华北落叶松**

学　名 *Larix gmelinii* var. *principis-rupprechtii*

松科
Pinaceae

落叶松属
Larix

[形态描述] 乔木，高达30米，胸径90厘米；树皮暗灰褐色，不规则纵裂，成小块片脱落；枝平展，具不规则细齿；树冠卵状圆锥形；一年生长枝较细，淡黄褐色，无毛或有短毛。叶倒披针状条形，先端尖或钝尖，上面中脉不隆起。苞鳞暗紫色，近带状矩圆形，仅球果基部苞鳞的先端露出；种子斜倒卵状椭圆形，灰白色，具不规则的褐色斑纹，种翅上部三角状，子叶5～7枚，下面无气孔线。花期3～5月，球果9月成熟。

[北京分布] 密云区（坡头，雾灵山）；门头沟区（小五台山，东灵山，百花山，小龙门）；房山区（白草畔，大安山）；延庆区（松山）；海淀区。

[保护策略] 减少人类干扰，增强自然更新能力；采集不同种群的种子，建立基因库；在气候变化影响较大的区域，引入适应性强的近缘落叶松物种，与华北落叶松形成混交林。

Morphological Description: Trees to 30 m tall; trunk to 90 cm d.b.h.; bark dark grayish-brown, irregularly longitudinally fissured, exfoliating in small irregular flakes; branches spreading, with irregular fine teeth; crown ovoid-conical; annual branchlets slender, pale yellowish-brown, glabrous or shortly pubescent. Leaves oblanceolate-linear, apex acute or obtuse-acuminate, midvein not raised adaxially. Seed scales dark purple, nearly banded-oblong, only apex of basal seed scales exposed on cones; seeds obliquely obovate-elliptic, grayish-white, with irregular brown streaks; seed wing triangular-ovate; cotyledons 5–7, without stomatal lines abaxially. Pollination Mar–May, seed maturity Sep.

Distribution in Beijing: Miyun District: Potou, Wuling Mountain; Mentougou District: Xiaowutai Mountain, Dongling Mountain, Baihua Mountain, Xiaolongmen; Fangshan District: Baicaopan, Da'an Mountain; Yanqing District: Songshan Mountain; Haidian District.

Protection Strategies: Reduce human interference to enhance natural regeneration capacity. Collect seeds from different populations to create a gene bank. In areas heavily affected by climate change, introduce highly adaptable related larch species to form mixed forests with *Larix gmelinii* var. *principis-rupprechtii*.

008

中文名 **白杆**

学　名 *Picea meyeri*

松科
Pinaceae

云杉属
Picea

[形态描述] 乔木，高达30米，胸径约60厘米；树皮灰褐色，裂成不规则的薄块片脱落；大枝近平展，树冠塔形；小枝有密生或疏生短毛或无毛。主枝之叶常辐射伸展，侧枝上面之叶伸展，两侧及下面之叶向上弯伸，四棱状条形，微弯曲。球果成熟前绿色，熟时褐黄色，矩圆状圆柱形；中部种鳞倒卵形，鳞背露出部分有条纹；种子倒卵圆形，种翅淡褐色。花期4月，球果9月下旬至10月上旬成熟。

[北京分布] 密云区（坡头）；怀柔区；门头沟区（东灵山，小龙门）；密云区（雾灵山）。

[保护策略] 防止水土流失，保护根系环境的稳定性；对山地和丘陵地区减少矿山开采等干扰活动；定期进行病虫害监测，引入病虫害的天敌昆虫；分析白杆在干旱、寒冷气候条件下的适应特性，开展白杆种植试验。

Morphological Description: Trees to 30 m tall; trunk to 60 cm d.b.h.; bark gray-brown, irregularly flaking; crown conical; branchlets yellow-brown, pubescent or glabrous. Leaves spreading radially, ascending on upper side of branchlets, spreading and curved upward on lower side, quadrangular-linear, slightly curved. Seed cones green, maturing brown-yellow, oblong-cylindric. Seed scales obovate, striate on exposed part abaxially. Seeds obovoid; wing pale brown. Pollination Apr, seed maturity Sep–Oct.

Distribution in Beijing: Miyun District: Potou; Huairou District; Mentougou District: Dongling Mountain, Xiaolongmen; Miyun District: Wuling Mountain.

Protection Strategies: Prevent soil erosion and maintain the stability of root environments. Reduce disruptive activities such as mining in mountainous and hilly areas. Conduct regular pest and disease monitoring and introduce natural predatory insects to manage outbreaks. Analyze the adaptability of *Picea meyeri* to arid and cold climates, and conduct planting trials for this species.

009

中文名 **青杆**

学　名 *Picea wilsonii*

松科
Pinaceae

云杉属
Picea

[形态描述] 乔木，高达50米，胸径达1.3米；树皮灰色或暗灰色，裂成不规则鳞状块片脱落；树冠塔形；一年生枝淡黄绿色或淡黄灰色，无毛，稀有疏生短毛，二三年生枝淡灰色、灰色或淡褐灰色。叶排列较密，在小枝上部向前伸展，小枝下面之叶向两侧伸展，四棱状条形，直或微弯。球果卵状圆柱形或圆柱状长卵圆形，黄褐色或淡褐色；中部种鳞倒卵形；种子倒卵圆形，种翅倒宽披针形，淡褐色。花期4月，球果10月成熟。

[北京分布] 密云区（坡头，南横岭，雾灵山）；门头沟区（江水河村，小龙门）；房山区（上方山）；延庆区（玉渡山）。

[保护策略] 限制山区开发活动，减少人为干扰对青杆林地的影响；在降雨集中期采取水土保持措施，降低地表径流和泥石流对青杆林地的冲刷威胁；建立病虫害定期监测机制，并优先采用生物防治技术；在适生区域推广青杆与近缘抗性树种的混交种植模式，提升林地生态系统的整体抗病虫害能力和抗逆性。

Morphological Description: Trees to 50 m tall; trunk to 1.3 m d.b.h.; bark gray or dark gray, irregularly flaking; crown pyramidal; one-year-old branchlets light yellowish green or light yellowish gray. glabrous rarely with sparse short hairs, two-to three-year-old branchlets light gray, gray, or light brown-gray. Leaves densely arranged, spreading forward on upper side of branchlets, spreading on lower side, quadrangular-linear, straight or slightly curved. Seed cones green, maturing yellow-brown or pale brown, ovoid-oblong. Seed scales at middle of cones obovate. Seeds obovoid; wing pale brown, oblanceolate. Pollination Apr, seed maturity Oct.

Distribution in Beijing: Miyun District: Potou, Nanhengling, Wuling Mountain; Mentougou District: Jiangshuihe Village, Xiaolongmen; Fangshan District: Shangfang Mountain; Yanqing District: Yudu Mountain.

Protection Strategies: Restrict development activities in mountainous areas to reduce the impact of human disturbance on *Picea wilsonii* forestlands. Implement soil and water conservation measures during concentrated rainfall periods to mitigate the scouring threat of surface runoff and debris flows to *Picea wilsonii* forestlands. Establish a regular monitoring mechanism for plant diseases and insect pests, and prioritize the adoption of biological control technologies. Promote mixed planting patterns of *Picea wilsonii* and related resistant tree species in suitable habitats to enhance the overall resistance of forest ecosystems to pests, diseases, and environmental stresses.

010

中文名 **杜松**
学　名 *Juniperus rigida*

柏科
Cupressaceae

刺柏属
Juniperus

@辰山注亚

[形态描述] 灌木或小乔木，高达10米；形成塔形或圆柱形的树冠；小枝下垂，幼枝三棱形。叶三叶轮生，条状刺形，质厚，坚硬，先端锐尖，槽内有1条窄白粉带，下面有明显的纵脊，雄球花椭圆状或近球状，球果圆球形，成熟前紫褐色，熟时淡褐黑色或蓝黑色，常被白粉；种子近卵圆形，有4条不显著的棱角。

[北京分布] 门头沟区（百花山，小龙门）。

[保护策略] 实施土壤保持措施，减少山地开发，改善土壤结构；引入天敌昆虫，定期使用生物杀菌剂防治锈病；在适宜土壤培育幼苗后移植，间隔种植以防湿气积聚，增加成活率。

Morphological Description: Shrubs erect, or small trees to 10 m; crown pyramidal or cylindric; branchlets pendulous, 3-angled when young. Leaves in whorls of 3, linear-needlelike, thick, deeply grooved with a narrow, white stomatal band adaxially, prominently keeled abaxially, apex sharply pointed. Pollen cones axillary, ellipsoid or subglobose. Seed cones axillary, light brownish blue or bluish black when ripe, usually glaucous, globose. Seeds often subovoid, indistinctly 4-ridged.

Distribution in Beijing: Mentougou District: Baihua Mountain, Xiaolongmen.

Protection Strategies: Implement soil conservation measures to reduce mountain development and improve soil structure. Introduce predatory insects and regularly apply biological fungicides to control rust disease. Cultivate seedlings in suitable soil conditions before transplanting, and plant with proper spacing to prevent moisture accumulation and increase survival rates.

011

中文名 **独角莲**

学　名 *Sauromatum giganteum*

天南星科
Araceae

斑龙芋属
Sauromatum

[形态描述] 块茎倒卵形、卵球形或卵状椭圆形。叶与花序同时抽出。叶柄圆柱形，密生紫色斑点；叶片幼时内卷如角状，后即展开，箭形，先端渐尖，基部箭状。佛焰苞紫色，管部圆筒形或长圆状卵形；檐部卵形，展开，先端渐尖常弯曲。肉穗花序几无梗，雌花序圆柱形；附属器紫色，圆柱形，直立，基部无柄，先端钝。雌花：子房圆柱形，顶部截平，胚珠2；柱头无柄，圆形。花期6~8月，果期7~9月。

[北京分布] 房山区（上方山）；昌平区（沟崖林场）。

[保护策略] 减少人为干扰对独角莲生境的影响；可采集种子进行人工迁地保护，并开展适地回归。

Morphological Description: Underground part a rhizome. Petiole green, with or without numerous purple spots; leaf blade ovate, cordate to hastate, apex acuminate. Inflorescence preceding or simultaneous with leaves. Spathe convolute at base, erect, oblong-ovate; limb erect or recurved, ovate. Spadix sessile or nearly so; female zone cylindric; ovary: basal part whitish, apical part purple, cylindric, apex truncate, 2-ovuled; stigma sessile, disciform; appendix sessile, cylindric, apex obtuse. Fl. Jun–Aug, fr. Jul–Sep.

Distribution in Beijing: Fangshan District: Shangfang Mountain; Changping District: Gou'ai Forest Farm.

Protection Strategies: Reduce the impact of human disturbance on the habitat of *Sauromatum giganteum*. Seeds may be collected for ex situ conservation, followed by reintroduction programs in suitable habitats.

012

中文名 **少花万寿竹**

学　名 *Disporum uniflorum*

秋水仙科
Colchicaceae

万寿竹属
Disporum

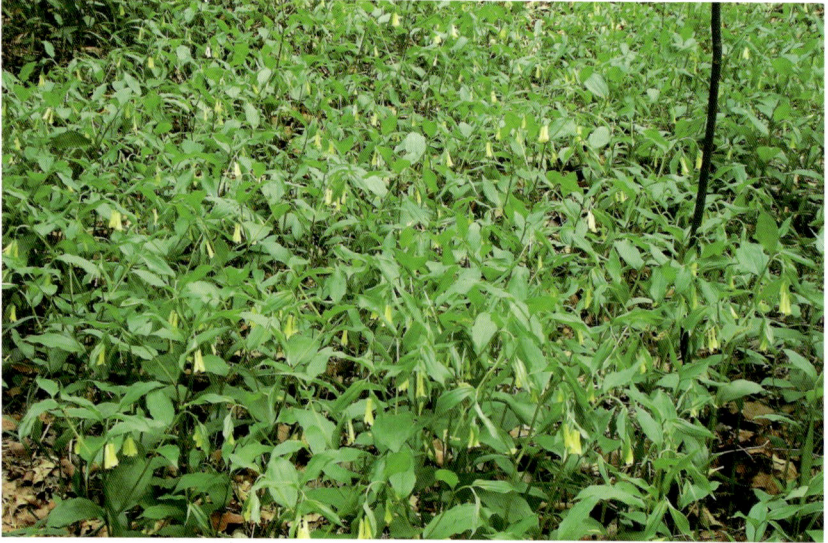

[形态描述] 根状茎肉质，横出。茎直立，上部具叉状分枝。叶薄纸质至纸质、矩圆形、卵形、椭圆形至披针形，下面色浅，脉上和边缘有乳头状突起，具横脉，先端骤尖或渐尖，基部圆形或宽楔形。花黄色、绿黄色或白色，1~3（~5）朵着生于分枝顶端；花被片近直出，倒卵状披针形，内面有细毛，边缘有乳头状突起，基部具短距；雄蕊内藏；花柱具3裂而外弯的柱头。浆果椭圆形或球形，具3颗种子；种子深棕色。花期3~6月，果期6~11月。

[北京分布] 房山区（上方山）。

[保护策略] 维持适宜湿度，减轻干燥气流对植物的不利影响；少花万寿竹的根部能够适应共生菌群的定植，可在原生境土壤中适量引入与其共生的根菌，以增强其根系吸收能力；定期排查常见病虫害，使用生物杀菌剂及释放捕食性昆虫进行防治。

Morphological Description: Rhizomes fleshy, horizontal. Stems erect, upper part with dichotomous branching. Leaves thinly papery to papery, oblong, ovate, elliptic, or

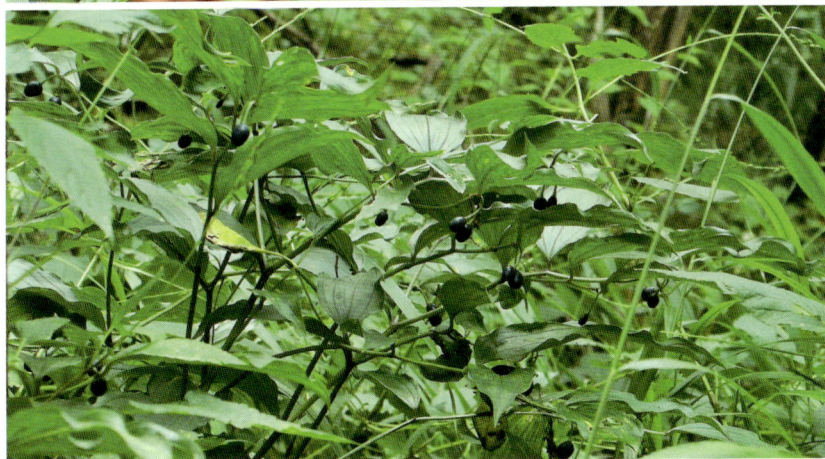

lanceolate, abaxially pale, papillose along midvein and margin, transverse veins present, apex abruptly acute or acuminate, base rounded or broadly cuneate. Flowers yellow, greenish-yellow, or white, 1–3(–5) borne at branch apex; perianth segments nearly straight, oblanceolate, adaxially minutely pubescent, margin papillose, base with a short spur; stamens included; styles with 3-lobed and recurved stigmas. Berries elliptic or globose, 3-seeded. Seeds dark brown. Fl. Mar.–Ju., fr. Ju.–Nov.

Distribution in Beijing: Fangshan District: Shangfang Mountain.

Protection Strategies: Maintain appropriate humidity to mitigate the adverse effects of dry air currents on plants. The roots of *Disporum uniflorum* adapt to symbiotic microbial communities, so introducing appropriate symbiotic root fungi into the planting soil can enhance root absorption. Regularly inspect common pests and diseases, use biological fungicides and release predatory insects for control.

013 　中文名 **宝珠草**　　　　秋水仙科
　　　 学　名 *Disporum viridescens*　Colchicaceae

　　　　　　　　　　　　　　　　　万寿竹属
　　　　　　　　　　　　　　　　　Disporum

[形态描述] 根状茎短，通常有长匍匐茎；根多而较细。茎高30～80厘米，有时分枝。叶椭圆形至卵状矩圆形，横脉明显，下面脉上和边缘稍粗糙。花淡绿色，1～2朵生于茎或枝的顶端；花被片张开，矩圆状披针形，脉纹明显，先端尖，基部囊状。浆果球形，黑色，有2～3颗种子。种子红褐色。花期5～6月，果期7～10月。

[北京分布] 昌平区（黑山寨）。

[保护策略] 保护林下及半阴环境，避免强光直射，维持适宜湿度，可种植于乔木或高大灌木的树荫下；定期监测，防止病害扩散，防治时尽量采用生态友好型杀菌剂；先在温室中育苗，再选择适应性强的幼苗移植至野外。

Morphological Description: Rhizome short, usually with long, creeping stolon; roots densely tufted. Stem often branched distally, 30–80 cm. Leaf blade ellitpic to ovate-oblong, cross veins indistinct, margin minutely scabrous. Inflorescences terminal, 1– or 2-flowered. Flowers widely opening. Tepals greenish white, oblong-lanceolate to lanceolate, 7-veined, base slightly saccate, apex long attenuate. Berries black, globose, 2- or 3-seeded. Seeds red-brown. Fl. May–Jun, fr. Jul–Oct.

Distribution in Beijing: Changping District: Heishanzhai.

Protection Strategies: It can be cultivated in protected forest understories or semi-shaded environments to avoid direct sunlight while maintaining appropriate humidity. Planting under tall trees or large shrubs for shade is ideal. Regular monitoring should be conducted to prevent the spread of diseases, using eco-friendly fungicides as needed. Seedlings can first be grown in a greenhouse, and robust, well-adapted young plants can then be transplanted into the wild.

014

中 文 名 **七筋姑**
学　　名 *Clintonia udensis*

百合科
Liliaceae

七筋姑属
Clintonia

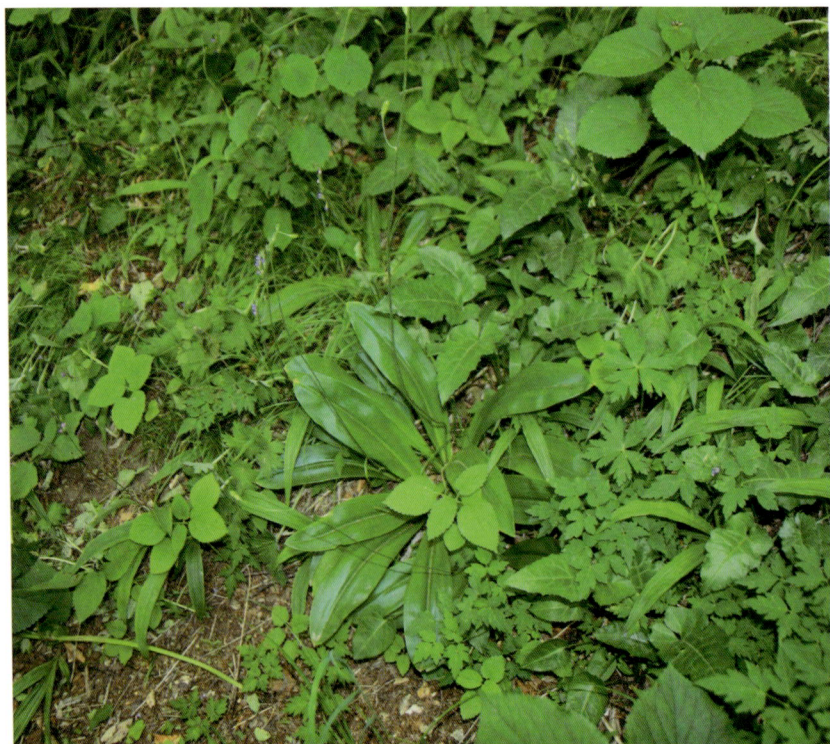

[形态描述] 根状茎较硬，粗约5毫米，有撕裂成纤维状的残存鞘叶。叶3～4枚，椭圆形、倒卵状矩圆形或倒披针形，幼时边缘有柔毛。花葶密生白色短柔毛；总状花序有花3～12朵，花梗密生柔毛；苞片早落；花白色，少有淡蓝色；花被片矩圆形，外面有微毛。果实球形至矩圆形；每室有种子6～12颗。花期5～6月，果期7～10月。

[北京分布] 门头沟区（百花山）；密云区（坡头）；房山区（白草畔，霞云岭）。

[保护策略] 保护七筋姑生长的湿润且排水良好的生境；七筋姑根系较为纤细，适合分株繁殖，可先在温室育苗，待根系稳定后进行分批移栽，以确保根系适应新土壤条件；针对叶部易染的真菌性病害，定期检查叶片状况。

Morphological Description: Rhizome stiff, ca. 5 mm in thick, covered with fibrous sheaths. Leaves 3–4, obovate, elliptic-obovate, or oblanceolate, margin pubescent when young. Scape densely white pubescent. Raceme 3–12-flowered; bracts caducous; pedicels densely pubescent. Tepals white or sometimes bluish, oblong, puberulent abaxially. Fruit globose to oblong. Seeds 6–12 per locule. Fl. May–Jun, fr. Jul–Oct.

Distribution in Beijing: Mentougou District: Baihua Mountain; Miyun District: Potou; Fangshan District: Baicaopan, Xiayunling.

Protection Strategies: Protect the moist and well-drained habitat where *Clintonia udensis* grows. Since *Clintonia udensis* has delicate roots, it is well-suited for propagation through division. Seedlings can be cultivated in a greenhouse first, and after root systems stabilize, they should be transplanted in batches to ensure proper adaptation to new soil conditions. Regularly inspect the leaves for fungal infections, which are common on the foliage, and take appropriate preventive measures.

015

中文名 **有斑百合**

学　名 *Lilium concolor* var. *pulchellum*

百合科
Liliaceae

百合属
Lilium

[形态描述] 鳞茎卵球形；鳞片卵形或卵状披针形，白色，少数近基部带紫色，有小乳头状突起。叶散生，条形，边缘有小乳头状突起。花1～5朵排成近伞形或总状花序；花直立，星状开展，深红色；花被片矩圆状披针形，蜜腺两边具乳头状突起，花被片有斑点；雄蕊向中心靠拢；花丝无毛；子房圆柱形。蒴果矩圆形。花期6～7月，果期8～9月。

[北京分布] 门头沟区（百花山，妙峰山，小龙门，东龙门涧，东灵山）；昌平区（莽山）；平谷区（千佛崖，金山）；密云区（遥桥峪村，冯家峪，雾灵山，石城镇）；延庆区（松山，兰角沟）；怀柔区（孙栅子）。

[保护策略] 保护生境，减少人为干扰和采摘；注意人工采种繁育，适时回归。

Morphological Description: Bulb ovoid; scales white, ovate or ovate-lanceolate. Stem occasionally tinged purple near base, papillose. Leaves scattered, linear, veins and margin papillose. Flowers 1–5 in a subumbel or raceme, erect. Tepals stellately spreading, deep red, spotted or unspotted, oblong-lanceolate to oblanceolate; nectaries papillose on both surfaces. Stamens converging; filaments glabrous. Capsule oblong. Fl. Jun–Jul, fr. Aug–Sep.

Distribution in Beijing: Mentougou District: Baihua Mountain, Miaofeng Mountain, Xiaolongmen, Donglongmenjian Canyon, Dongling Mountain; Changping District: Mangshan Mountain; Pinggu District: Qianfo Cliff, Jinshan Mountain; Miyun District: Yaoqiaoyu Village, Fengjiayu, Wuling Mountain, Shicheng Town; Yanqing District: Songshan Mountain, Lanjiaogou; Huairou District: Sunzhazi.

Protection Strategies: Protect habitats by reducing human disturbance and picking. Implement ex situ conservation through artificial seed collection, followed by strategic reintroduction.

016

中文名 **山丹**
学　名 *Lilium pumilum*

百合科
Liliaceae

百合属
Lilium

[形态描述] 鳞茎卵形或圆锥形；鳞片矩圆形或长卵形，白色。茎高15～60厘米，有小乳头状突起，有的带紫色条纹。叶散生于茎中部，条形，中脉下面突出，边缘有乳头状突起。花单生或数朵排成总状花序，鲜红色，通常无斑点，下垂；花被片反卷，蜜腺两边有乳头状突起；花丝无毛，花药黄色；子房圆柱形。蒴果矩圆形。花期7～8月，果期9～10月。

[北京分布] 房山区（周口店，圣水峪，罗家峪，浦洼，十渡，云水洞，云居寺，上方山）；延庆区（松山，兰角沟，八达岭，玉渡山，海坨山）；昌平区（莽山）；门头沟区（东灵山，妙峰山，小龙门，百花山）；海淀区（西山，香山，卧佛寺）；密云区（遥桥峪）；怀柔区（喇叭沟门，琉璃庙）；平谷区（金山）。

[保护策略] 保护生境，减少人为干扰和采摘；注意人工采种繁育，适时回归。

Morphological Description: Bulb ovoid or conical; scales white, oblong or narrowly ovate. Stem sometimes streaked with purple, 15–60 cm, papillose. Leaves scattered near middle of stem, linear, midvein prominent abaxially, margin papillose. Flowers solitary or several in a raceme, nodding. Tepals revolute, bright red, usually unspotted, not minutely papillose adaxially; nectaries papillose on both surfaces. Filament glabrous; anthers yellow. Capsule oblong. Fl. Jul–Aug, fr. Sep–Oct.

Distribution in Beijing: Fangshan District: Zhoukoudian, Shengshuiyu, Luojiayu, Puwa, Shidu, Yunshuidong, Yunju Temple, Shangfang Mountain; Yanqing District: Songshan Mountain, Lanjiaogou, Badaling, Yudu Mountain, Haituo Mountain; Changping District: Mangshan Mountain; Mentougou District: Dongling Mountain, Miaofeng Mountain, Xiaolongmen, Baihua Mountain; Haidian District: Xishan, Xiangshan, Wofo Temple; Miyun District: Yaoqiaoyu; Huairou District: Labagoumen, Liulimiao; Pinggu District: Jinshan Mountain.

Protection Strategies: Protect habitats by reducing human disturbance and picking. Implement ex situ conservation through artificial seed collection, followed by strategic reintroduction.

017 中文名 **菰（gū）**
学　名 *Zizania latifolia*

禾本科
Poaceae

菰属
Zizania

[形态描述] 多年生，具匍匐根状茎。秆高大直立，高1～2米，基部节上生不定根。叶鞘长于其节间，肥厚，有小横脉；叶舌膜质，顶端尖；叶片扁平宽大。圆锥花序，分枝多数簇生；雄小穗两侧压扁，着生于花序下部或分枝之上部，带紫色，外稃具5脉，内稃具3脉，中脉成脊，具毛，雄蕊6枚；雌小穗圆筒形，外稃之5脉粗糙，芒长20～30毫米，内稃具3脉。颖果圆柱形。

[北京分布] 海淀区（翠湖湿地）；大兴区；延庆区（野鸭湖，张山营）。

[保护策略] 保持水体清洁，种植伴生水生植物如芦苇等吸收多余养分，维护水体生态平衡；防治蚜虫、螟虫等常见害虫。

Morphological Description: Perennial, with creeping rhizomes. Culms tall and erect, 1–2 m high, with adventitious roots arising from the nodes at the base. Leaf sheaths longer than their internodes, thick and fleshy, with small cross veins; ligule membranous, apex acute;leaf blades flat and broad. Panicle with numerous branches clustered; male spikelets laterally compressed, borne in the lower part of the inflorescence or on the upper part of the branches, purple tinged, lemma with 5 veins, palea with 3 veins, the midvein forming a ridge, hairy, stamens 6; female spikelets cylindrical, lemma of female spikelets with 5 rough veins, awn 20–30 mm long; palea with 3 veins. Caryopsis cylindrical.

Distribution in Beijing: Haidian District: Cuihu Wetland; Daxing District; Yanqing District: Wild Duck Lake, Zhangshanying.

Protection Strategies: Ensure water quality remains unpolluted by controlling nutrient loads. Introduce companion aquatic plants, such as reeds, to absorb excess nutrients and maintain ecological balance in the water. Regularly monitor and manage common pests, including aphids and stem borers, using integrated pest management techniques.

018

中 文 名 **房山紫堇**

学　　名 *Corydalis fangshanensis*

罂粟科
Papaveraceae

紫堇属
Corydalis

[形态描述] 灰绿色丛生草本，高20～30厘米，具主根。茎不分枝，叶柄稍长于叶片；叶片披针形，二回羽状全裂，末回羽片倒卵形，基部楔形，三深裂。总状花序疏具多花；苞片披针形，约与花梗等长；花淡红紫色至近白色；萼片卵圆形；外花瓣宽展，具浅鸡冠状突起；距囊状；爪约与瓣片等长；内花瓣具高鸡冠状突起。蒴果线形，下垂；种子肾形，种阜柄状，紧贴。

[北京分布] 房山区（上方山，圣水峪村，霞云岭村，十渡）；门头沟区（龙门涧）。

[保护策略] 选择疏松透气、富含腐殖质的土壤区域栽植，避免积水；用灌木或草本植物伴生种植，提供适度遮阴；先在温室育苗，再将成苗移植至野外半阴环境；防治蚜虫及叶部真菌病害。

Morphological Description: Herbs, glaucous, 20–30 cm tall, with taproot. Stems much branched with condensed lower internodes. Petiole ca. as long as blade; blade ovate-lanceolate, bipinnate; ultimate pinnae obovate, base cuneate, deeply 3-divided. Raceme stalked, 7-15(-20)-flowered; bracts lanceolate, ca. equal to pedicels. Flowers white, apex of flower either purplish or white with greenish keels. Sepals orbicular. Outer petals with broad reflexed margins at apex, very narrowly crested; spur saccate; inner petals with broad crest clearly overtopping apex; claw shorter than limb. Capsule drooping, linear. Seeds in 1 row, reniform; caruncle tightly adhering.

Distribution in Beijing: Fangshan District: Shangfang Mountain, Shengshuiyu Village, Xiayunling Village, Shidu; Mentougou District: Longmenjian Canyon.

Protection Strategies: Select a well-drained, breathable soil rich in humus for planting, avoiding waterlogging. Plant alongside shrubs or herbaceous plants to provide moderate shade. Begin with greenhouse seedling cultivation, and once the seedlings are established, transplant them to a semi-shaded outdoor environment. Implement pest control measures to prevent aphids and fungal leaf diseases.

019

中文名 **红毛七**
学　名 *Caulophyllum robustum*

小檗科
Berberidaceae

红毛七属
Caulophyllum

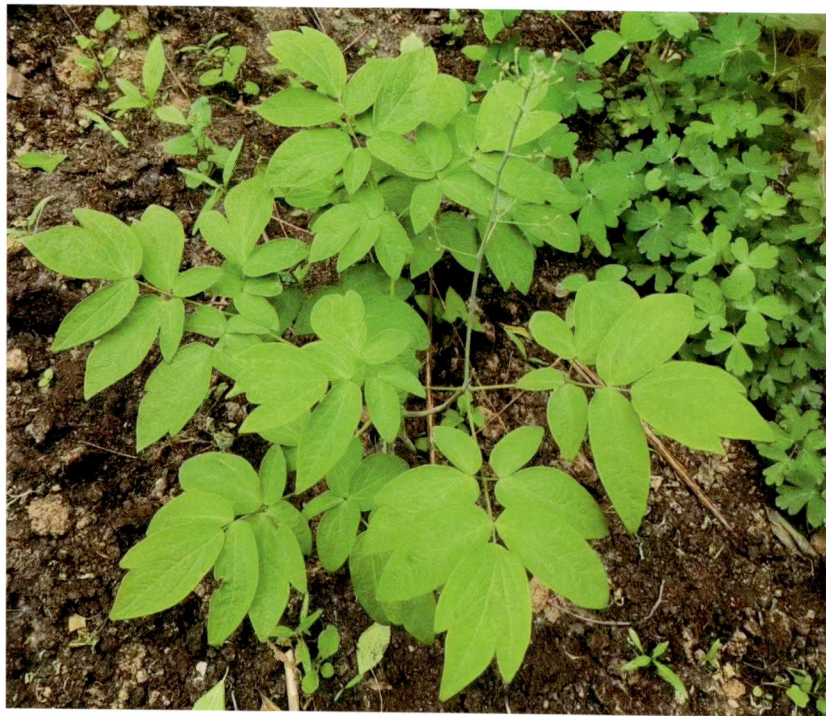

[形态描述] 多年生草本，植株高达80厘米。根状茎粗短。茎生2叶，互生；小叶卵形、长圆形或阔披针形，先端渐尖，基部宽楔形，全缘，有时2～3裂，上面绿色，背面淡绿色，两面无毛；顶生小叶具柄，侧生小叶近无柄。花淡黄色；萼片倒卵形；花瓣基部缢缩呈爪；雄蕊花丝稍长于花药；子房具2枚基生胚珠，花后子房开裂，露出2枚球形种子。种子微被白粉，熟后蓝黑色，外被肉质假种皮。

[北京分布] 怀柔区（黑坨山）；延庆区；密云区；门头沟区；平谷区；房山区。

[保护策略] 保护生境；在地表覆盖稻草或树叶，减少水分蒸发；搭建遮阳网或在周围种植乔灌木以防止阳光直射。

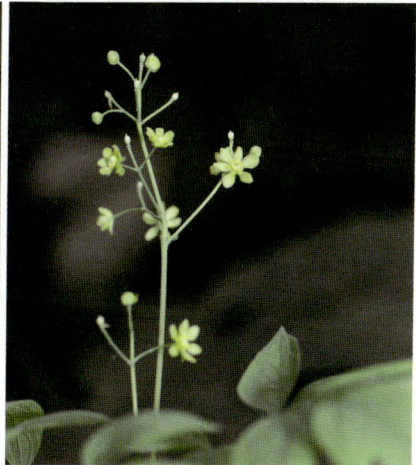

Morphological Description: Herbs, to 80 cm tall. Rhizomes short, stout. Stem leaves 2; terminal leaflet usually petiolulate, lateral leaflets subsessile; leaflets abaxially pale green or grayish white, adaxially green, ovate, oblong, or broadly lanceolate, both surfaces glabrous, base broadly cuneate, margin entire, sometimes 2- or 3-lobed, apex acuminate. Flowers pale yellow. Sepals obovate. Petals base clawed. Stamens, filaments longer than anthers. ovules 2. Seeds naked at maturity; seed coat blue, fleshy, glaucous.

Distribution in Beijing: Huairou District: Heituo Mountain; Yanqing District; Miyun District; Mentougou District; Pinggu District; Fangshan District.

Protection Strategies: Protect habitats. Mulch the surface with straw or leaves to reduce water evaporation. Use shade nets or plant trees and shrubs around to prevent direct sunlight.

020

中文名 **草芍药**

学　名 *Paeonia obovata*

芍药科
Paeoniaceae

芍药属
Paeonia

[形态描述] 多年生草本。根粗壮。茎高30~70厘米，无毛。茎下部叶为二回三出复叶；顶生小叶倒卵形，顶端短尖，基部楔形，全缘；侧生小叶比顶生小叶小；茎上部叶为三出复叶或单叶。单花顶生；萼片3–5，宽卵形；花瓣白色、红色、紫红色，倒卵形；花药长圆形；花盘浅杯状；心皮2~3，无毛。蓇葖卵圆形，成熟时果皮反卷呈红色。花期5~6月；果期9月。

[北京分布] 门头沟区（百花山，小龙门，清水镇，东灵山）；密云区（坡头，云蒙山，雾灵山，遥桥峪）；怀柔区（喇叭沟门）；延庆区（海坨山，松山）；房山区（霞云岭）。

[保护策略] 草芍药适合在半阴、温凉的山地林下生长，土壤要求疏松、排水良好且富含腐殖质。应选择朝北或东南坡地回归种植，避免强光直射，以保持湿润环境。

Morphological Description: Herbs perennial, 30–70 cm tall. Roots thick. Stems glabrous, 2-ternate; leaflets obovate, base cuneate, margin entire, apex rounded or acute. Flowers solitary, terminal, single. Sepals 3–5, apex mostly rounded. Petals white, rose, pink-red, red, purple-red, or rarely white with a pinkish base or margin, obovate. Disc yellow, annular. Carpels 2–3, glabrous. Follicles gradually recurved, ellipsoid. Fl. May–Jun, fr. Sep.

Distribution in Beijing: Mentougou District: Baihua Mountain, Xiaolongmen, Qingshui Town, Dongling Mountain; Miyun District: Potou, Yunmeng Mountain, Wuling Mountain, Yaoqiaoyu; Huairou District: Labagoumen; Yanqing District: Haituo Mountain, Songshan Mountain; Fangshan District: Xiayunling.

Protection Strategies: *Paeonia obovata* thrives in semi-shaded, cool mountain forest environments. It prefers loose, well-drained soil that is rich in humus. It is advisable to choose north-facing or southeast-facing slopes for reintroduction planting to avoid direct strong sunlight and maintain a moist environment.

021

中文名 **辽吉侧金盏花**
学　名 *Adonis ramosa*

毛茛科
Ranunculaceae

侧金盏花属
Adonis

[形态描述] 多年生草本。茎无毛或顶部有稀疏短柔毛，下部或上部分枝。基部和下部叶鳞片状，卵形或披针形。茎中部以上叶约4，无柄或近无柄；叶片宽菱形，二至三回羽状全裂，末回裂片披针形或线状披针形。花单生茎或枝的顶端；萼片约5，灰紫色，宽卵形；全缘或上部边缘有1～2小齿；花瓣约13，黄色，长圆状倒披针形；雄蕊花药长圆形；心皮近无毛。3～4月开花。

[北京分布] 延庆区（四海）。

[保护策略] 避免对其自然栖息地的人为开发和破坏；定期对辽吉侧金盏花种群进行调查和监测；进行外来物种的控制和清除，确保辽吉侧金盏花的生长环境免受竞争威胁。

Morphological Description: Plants perennial. Stems, glabrous or apically sparsely pubescent, branched. Basal leaves and lower stem leaves scalelike, ovate to lanceolate. Leaves apically on stem ca. 4, sessile to subsessile; leaf blade broadly rhombic, 2 or 3 pinnately divided; ultimate segments linear to linear-lanceolate. Sepals ca. 5, gray-purple, broadly ovate, margin entire, and apically with 1 or 2 small teeth. Petals ca. 13, yellow, oblong-oblanceolate. Anthers oblong. Pistils subglabrous. Fl. Mar–Apr.

Distribution in Beijing: Yanqing District: Sihai.

Protection Strategies: Avoid human activities that lead to the development and destruction of its natural habitat. Regularly survey and monitor the populations of *Adonis ramosa*. Control and eliminate invasive species to ensure that the growth environment of *Adonis ramosa* remains unaffected by competitive pressures.

022

中文名 **多被银莲花**
学　名 *Anemone raddeana*

毛茛科
Ranunculaceae

银莲花属
Anemone

[形态描述] 植株高 10～30 厘米。根状茎横走。基生叶 1，有长柄；叶片三全裂，叶柄有疏柔毛。苞片 3，叶片近扇形，三全裂，中全裂片倒卵形或倒卵状长圆形；萼片 9～15，白色，长圆形或线状长圆形，无毛；花药椭圆形；子房密被短柔毛。4～5 月开花。

[北京分布] 房山区（蒲洼）。

[保护策略] 进行小规模的人工繁殖，逐步引入适宜的林下阴湿环境中；保持半阴、疏松的土壤和较稳定的水分条件；对自然灾害造成土壤侵蚀和水源破坏的栖息地进行必要的恢复工程。

Morphological Description: The plant grows 10–30 cm in height. Rhizome creeping horizontally. Basal leaf solitary, long-petiolate; lamina ternately fully divided, petiole sparsely pubescent. Bracts 3, subflabellate in outline, tripartite with median segment obovate or obovate-oblong. Sepals 9–15, white, oblong or linear-oblong, glabrous. Anthers ellipsoid. Ovary densely covered with short pubescence. Fl. Apr–May.

Distribution in Beijing: Fangshan District: Puwa.

Protection Strategies: Conduct small-scale artificial propagation and gradually introduce the plants into suitable shady and moist forest environments. Maintain semi-shaded conditions, loose soil, and stable moisture levels. Implement necessary restoration projects in habitats affected by soil erosion and water source damage caused by natural disasters.

023

中文名 **金莲花**

学　名 *Trollius chinensis*

毛茛科
Ranunculaceae

金莲花属
Trollius

[形态描述] 植株全体无毛。茎高30～70厘米，不分枝。基生叶1～4个；叶片五角形，基部心形，三全裂；叶柄基部具狭鞘。茎生叶似基生叶。花单独顶生或成聚伞花序；萼片10～15片，椭圆状倒卵形或倒卵形，顶端圆形；花瓣18～21个，稍长于萼片或与萼片近等长，狭线形；心皮20～30。种子近倒卵球形，黑色光滑。6～7月开花，8～9月结果。

[北京分布] 门头沟区（百花山，东灵山，小龙门）；延庆区（海坨山，松山）；房山区（蒲洼，霞云岭）；密云区（坡头）；怀柔区（喇叭沟门，孙栅子）。

[保护策略] 定期清除或控制其栖息地内的外来入侵植物；在其生长区域设置围栏和警示牌，防止人为踩踏和采摘。

Morphological Description: Plant entirely glabrous. Stem 30–70 cm tall, unbranched. Basal leaves 1–4; lamina pentagonal, base cordate, ternately fully divided; petiole base narrowly sheathing. Cauline leaves similar to basal ones. Flowers solitary and terminal or arranged in a cyme; sepals 10–15, elliptic-obovate or obovate, apex rounded; petals 18–21, slightly longer than sepals or nearly equal in length, narrowly linear; carpels 20–30; seeds nearly obovoid, black and smooth. Fl. Jun–Jul, fr. Aug–Sep.

Distribution in Beijing: Mentougou District: Baihua Mountain, Dongling Mountain, Xiaolongmen; Yanqing District: Haituo Mountain, Songshan Mountain; Fangshan District: Puwa, Xiayunling; Miyun District: Potou; Huairou District: Labagoumen, Sunzhazi.

Protection Strategies: Regularly remove or control invasive plant species within the habitat to maintain ecological balance. Install fencing and warning signs around the growth area to prevent trampling and unauthorized picking by humans.

024

中文名 **小丛红景天**

学　名 *Rhodiola dumulosa*

景天科
Crassulaceae

红景天属
Rhodiola

[形态描述] 多年生草本。根颈粗壮，地上部分常被有残留的老枝。花茎聚生主轴顶端，直立或弯曲，不分枝。叶互生，线形至宽线形，全缘。花序聚伞状，有 4~7 花；萼片 5，线状披针形；花瓣 5，白或红色，披针状长圆形，先端渐尖；雄蕊 10，较花瓣短；心皮 5，卵状长圆形。种子长圆形，有微乳头状突起，有狭翅。花期 6~7 月，果期 8 月。

[北京分布] 门头沟区（东灵山，百花山，小龙门）；房山区（白草畔）；延庆区（松山，海坨山）。

[保护策略] 避免在其栖息地附近进行登山开发和建设活动；定期清理和监控入侵植物，避免其生态位被外来物种取代；对其种群分布和生态特性进行长期监测。

Morphological Description: Caudex branched, robust; persistent old flowering stems present as remnants. Flowering stems aggregated apically on caudex, simple, erect or curved. Stem leaves alternate, linear to broadly so, margin entire. Inflorescences cymose, 4-7-flowered. Flowers unequally 5-merous. Sepals linear-lanceolate. Petals erect, white or red, lanceolate-oblong, apex acuminate and long mucronate. Stamens 10, shorter than petals. Carpels erect, ovoid-oblong. Seeds oblong, finely mammillate, narrowly winged. Fl. Jun-Jul, fr. Aug.

Distribution in Beijing: Mentougou District: Dongling Mountain, Baihua Mountain, Xiaolongmen; Fangshan District: Baicaopan; Yanqing District: Songshan Mountain, Haituo Mountain.

Protection Strategies: Avoid conducting hiking development and construction activities near its habitat. Regularly clear and monitor invasive plants to prevent its ecological niche from being replaced by alien species. Conduct long-term monitoring of its population distribution and ecological characteristics.

025

中文名 **狭叶红景天**
学　名 *Rhodiola kirilowii*

景天科
Crassulaceae

红景天属
Rhodiola

[形态描述] 多年生草本。根粗，直立。根颈先端被三角形鳞片。叶密，互生，线形至线状披针形，边缘有疏锯齿，或有时全缘。花序伞房状；雌雄异株；萼片三角形；花瓣5或4，绿黄色，倒披针形；雄花中雄蕊10或8，与花瓣同长或稍超出，花丝花药黄色；鳞片5或4，近正方形或长方形；心皮5或4，直立。蓇葖果披针形；种子长圆状披针形。

[北京分布] 门头沟区（百花山，东灵山，小龙门）；房山区（霞云岭）；延庆区（松山，海坨山）。

[保护策略] 维持其生境良好的排水、光照和较冷凉的温度；在其生长区域建立隔离区，尽量避免对其生长环境的干扰；采取土壤覆盖和坡面稳定技术以防止土壤侵蚀。

Morphological Description: Roots erect, thick. Caudex apex covered with triangular scales. Leaves dense, alternate, linear to linear-lanceolate, margin sparsely serrulate, or sometimes entire. Inflorescence corymbose; plants dioecious. Sepals triangular; petals 5 or 4, greenish yellow, oblanceolate. In male flowers, stamens 10 or 8, equaling or slightly longer than petals; filaments and anthers yellow; scales 5 or 4, subquadrangular or rectangular. Carpels 5 or 4, erect. Follicles lanceolate; seeds blong-lanceolate.

Distribution in Beijing: Mentougou District: Baihua Mountain, Dongling Mountain, Xiaolongmen; Fangshan District: Xiayunling; Yanqing District: Songshan Mountain, Haituo Mountain.

Protection Strategies: Maintain proper drainage, light conditions, and a relatively cool temperature in its habitat. Establish buffer zones around its growth areas to minimize disturbances to its environment. Implement soil cover and slope stabilization techniques to prevent erosion.

026

中 文 名 **黄芪**

学　　名 *Astragalus membranaceus*

豆科
Fabaceae

黄芪属
Astragalus

[形态描述] 多年生草本。主根肥厚，木质，常分枝。茎直立，上部多分枝，有细棱，被白色柔毛。羽状复叶有13～27片小叶；托叶离生，卵形、披针形或线状披针形，下面被白色柔毛或近无毛；小叶椭圆形或长圆状卵形，下面被伏贴白色柔毛。总状花序稍密；苞片线状披针形，背面被白色柔毛；小苞片2；花萼钟状，外面被白色或黑色柔毛，三角形至钻形；花冠黄色或淡黄色，旗瓣倒卵形；子房有柄，被细柔毛。荚果薄膜质，半椭圆形，顶端具刺尖，无毛；种子3～8颗。花期6～8月，果期7～9月。

[北京分布] 门头沟区（百花山）。

[保护策略] 维持生境土壤的疏松性和透气性；推广人工繁殖技术；定期监测，及时清除其栖息地的外来入侵植物。

Morphological Description: Perennial herb. Taproot thick, woody, often branched. Stem erect, extensively branched in the upper part, with fine ridges and covered in white pubescence. Leaves pinnately compound, with 13–27 leaflets; stipules free, ovate, lanceolate, or linear-lanceolate, white pubescent below or nearly glabrous; leaflets elliptical or oblong-ovate, adaxially glabrous and abaxially appressed with white pubescence. Racemes moderately dense; bracts linear-lanceolate, white pubescent on the abaxial side; bracteoles 2; calyx campanulate, externally covered in white or black pubescence, with triangular to narrowly triangular teeth; corolla yellow or pale yellow, standard petal obovate; ovary stipitate, covered in fine pubescence. Pods membranous, semi-elliptical, with a pointed apex, glabrous, containing 3–8 seeds. Fl. Jun–Aug, fr. Jul–Sep.

Distribution in Beijing: Mentougou District: Baihua Mountain.

Protection Strategies: Maintain the looseness and aeration of the habitat soil. Promote artificial propagation techniques. Conduct regular monitoring and promptly remove invasive alien plants from its habitat.

027

中文名 **齿叶白鹃梅**

学　名 *Exochorda serratifolia*

蔷薇科
Rosaceae

白鹃梅属
Exochorda

[形态描述] 落叶灌木，高达2米；冬芽卵形，先端圆钝。叶片椭圆形或长圆状倒卵形，基部楔形或宽楔形，中部以上有锐锯齿，下面全缘。总状花序，有花4~7朵，无毛；萼筒浅钟状，无毛；萼片三角卵形，先端急尖，全缘无毛；花瓣长圆形至倒卵形，先端微凹，基部有长爪，白色；雄蕊25。花期5~6月，果期7~8月。

[北京分布] 延庆区（千家店镇，珍珠泉乡）；怀柔区（汤河口镇，杏树台村，温栅子村，琉璃庙）；密云区（花园村）。

[保护策略] 在其生长区域周围设立缓冲区或围栏，防止人为踩踏、采摘或砍伐；建立长期监测系统。

Morphological Description: Shrubs to 2 m tall; buds purple-red, ovoid, apex obtuse. Leaf blade elliptic to oblong-ovate, base cuneate or broadly so, margin entire below middle, serrate above middle, apex obtuse or acute. Raceme 4–7-flowered, glabrous. Hypanthium shallowly campanulate, glabrous. Sepals triangular, base long clawed, apex emarginate. Stamens ca. 25. Fl. May–Jun, fr. Jul–Aug.

Distribution in Beijing: Yanqing District: Qianjiadian Town, Zhenzhuquan Township; Huairou District: Tanghekou Town, Xingshutai Village, Wenzhazi Village, Liulimiao; Miyun District: Huayuan Village.

Protection Strategies: Set up buffer zones or fences around its habitat to prevent trampling, harvesting, or logging. Establish a long-term monitoring system.

028 中文名 **水榆花楸**
学　名 *Sorbus alnifolia*

蔷薇科
Rosaceae

花楸属
Sorbus

[形态描述] 乔木，高达20米；小枝圆柱形，具灰白色皮孔，幼时微具柔毛；冬芽卵形，外具数枚暗红褐色无毛鳞片。叶片卵形至椭圆卵形，边缘有不整齐的尖锐重锯齿，上下两面无毛。复伞房花序较疏松，具花6～25朵；萼筒钟状；萼片三角形，先端急尖；花瓣卵形或近圆形，白色；雄蕊20，短于花瓣；花柱2，光滑无毛。果实椭圆形或卵形，红色或黄色。花期5月，果期8～9月。

[北京分布] 怀柔区（怀北镇，百泉山，渤海镇三岔村）；昌平区（南口镇）；密云区（白道峪，云蒙山，古北口）；延庆区（八达岭林场，三堡村，大庄科乡，兰角沟，松山）；房山区（圣水峪）。

[保护策略] 确保其生境的水源充足，防止土壤干燥影响其生长；加强水土保持，防止河岸侵蚀，维持其生境的土壤湿润性和稳定性；在其分布区域设立隔离带。

Morphological Description: Trees to 20 m tall. Branchlet terete, puberulent when young, with white lenticels; buds ovoid; scales several, dark reddish brown, glabrous. Leaf blade ovate to elliptic-ovate or suborbicular, both surfaces glabrous, margin irregularly sharply doubly serrate or lobed. Compound corymbs terminal, loosely 6–25-flowered. Hypanthium campanulate. Sepals triangular, apex acute. Petals white, suborbicular to oblong-ovate. Stamens 20; filaments white, slightly shorter than petals. Styles 2, glabrous. Fruit red, oblong, globose, sepals caducous, leaving a small annular scar. Fl. May, fr. Aug–Sep.

Distribution in Beijing: Huairou District: Huaibei Town, Baiquan Mountain, Sancha Village in Bohai Town; Changping District: Nankou Town; Miyun District: Baidaoyu, Yunmeng Mountain, Gubeikou; Yanqing District: Badaling Forest Farm, Sanbu Village, Dazhuangke Township, Lanjiaogou, Songshan Mountain; Fangshan District: Shengshuiyu.

Protection Strategies: Ensure a sufficient water supply in its habitat to prevent soil dryness from affecting its growth. Strengthen soil and water conservation efforts and prevent riverbank erosion to maintain soil moisture and stability in its habitat. Establish buffer zones in its distribution areas.

029

中文名 **北京花楸**

学　名 *Sorbus discolor*

蔷薇科
Rosaceae

花楸属
Sorbus

[形态描述]　乔木，高达10米；小枝圆柱形，嫩枝无毛；冬芽长圆卵形，外被数枚棕褐色鳞片。奇数羽状复叶；托叶宿存。复伞房花序较疏松，有多数花朵，总花梗和花梗均无毛；萼筒钟状，内外两面均无毛；萼片三角形；花瓣卵形或长圆卵形，白色；雄蕊15～20；花柱3～4。果实卵形，白色或黄色，先端具宿存闭合萼片。花期5月，果期8～9月。

[北京分布]　门头沟区（小龙门，百花山，齐家庄，东灵山）；昌平区（老峪沟村）；房山区（上方山，霞云岭）；怀柔区（老人沟）；延庆区（玉渡山，松山，海坨山）。

[保护策略]　确保其分布区域的水分供应，通过维护周边水源、修建小型蓄水池等措施，保持土壤湿润；定期巡查其栖息地，及时清除入侵植物；评估极端天气或气候变化对北京花楸的影响。

Morphological Description: Trees to 10 m tall. Branchlets terete, glabrous or nearly so when young; buds oblong-ovoid; scales several, brownish. Leaves imparipinnate; stipules persistent. Inflorescences loose, many flowered; rachis and pedicels glabrous. Hypanthium campanulate, abaxially glabrous. Sepals triangular. Petals white, ovate or oblong-ovate. Stamens 15–20. Styles 3 or 4. Fruit white or yellow, ovoid; sepals persistent. Fl. May, fr. Aug–Sep.

Distribution in Beijing: Mentougou District: Xiaolongmen, Baihua Mountain, Qijiazhuang, Dongling Mountain; Changping District: Laoyugou Village; Fangshan District: Shangfang Mountain, Xiayunling; Huairou District: Laorengou; Yanqing District: Yudu Mountain, Songshan Mountain, Haituo Mountain.

Protection Strategies: Ensure adequate water supply in its distribution areas by maintaining surrounding water sources and constructing small reservoirs to keep the soil moist. Regularly inspect its habitat to promptly remove invasive plants. Evaluate the impact of extreme weather or climate change on *Sorbus discolor*.

030 中文名 **北枳椇**
学　名 *Hovenia dulcis*

鼠李科
Rhamnaceae

枳椇属
Hovenia

[形态描述] 高大乔木，稀灌木，高达10余米；小枝褐色或黑紫色，无毛，有不明显的皮孔。叶卵圆形、宽矩圆形或椭圆状卵形，边缘有不整齐的锯齿或粗锯齿。花黄绿色，顶生，稀兼腋生的聚伞圆锥花序；萼片卵状三角形；花瓣倒卵状匙形，向下渐狭成爪部；子房球形，花柱3浅裂；花序轴结果时稍膨大。种子深栗色或黑紫色。花期5～7月，果期8～10月。

[北京分布] 房山区（上方山）；昌平区（沟崖林场）。

[保护策略] 保护其分布区的水源，特别是在干旱季节，要提供适量的水分补给，保持栖息地土壤湿润，并防止水土流失；定期巡查，及时清除生境中的入侵植物；根据环境变化的趋势及时调整保护措施。

Morphological Description: Trees, rarely shrubs, deciduous, to 10 m tall. Branchlets brown or black-purple, glabrous, with inconspicuous lenticels. Leaf blade ovate, broadly oblong, or elliptic-ovate, margin irregularly serrate or coarsely serrate. Flowers yellow-green, in terminal, or rarely axillary, asymmetrical cymose panicles. Sepals ovate-triangular. Petals clawed, obovate-spatulate. Ovary globose; style shortly 3-fid; peduncles and pedicels becoming fleshy and juicy at fruit maturity. Seeds deep brown or black-purple. Fl. May–Jul, fr. Aug–Oct.

Distribution in Beijing: Fangshan District: Shangfang Mountain; Changping District: Gou'ai Forest Farm.

Protection Strategies: Protect the water sources in its distribution area, especially by providing appropriate water supply during dry seasons to maintain moist habitat soil and prevent soil erosion. Conduct regular inspections and promptly remove invasive plants from the habitat. Adjust conservation measures in response to environmental changes.

031

中文名 **脱皮榆**
学　名 *Ulmus lamellosa*

榆科
Ulmaceae

榆属
Ulmus

[形态描述] 落叶小乔木，高 8～12 米；树皮灰色或灰白色，不断地裂成不规则薄片脱落；小枝上无扁平而对生的木栓翅。叶倒卵形，密生硬毛。花常自混合芽抽出，春季与叶同时开放。翅果常散生于新枝的近基部，圆形至近圆形，两面及边缘有密毛，顶端凹，果核位于翅果的中部；宿存花被钟状，被短毛；果梗密生伸展的腺状毛与柔毛。

[北京分布] 昌平区（虎峪、沟崖）；门头沟区（龙门涧，长条峪，燕家台，南石洋峡谷）；房山区（上方山，蒲洼）；延庆区（西大庄科村）。

[保护策略] 保护其生长区域的土壤结构，避免土壤板结和水分过多导致根系腐烂；在其分布区域设立保护区域，尽量减少人为干扰；开展人工繁殖研究，通过科学的方法提高种子发芽率和幼苗存活率，逐步扩大种群规模。

Morphological Description: Trees, 8–12 m tall, deciduous. Bark gray to grayish white, exfoliating in irregular flakes. Branchlets sometimes with a corky layer. Leaf blade obovate, abaxially scabrous and densely pubescent when young, adaxially scabrous and densely hirsute or with trichome scars, base oblique, margin simply or doubly serrate with blunt teeth, apex caudate to cuspidate. Flowers from mixed buds, appearing at same time as leaves. Samara usually scattered near base of branchlets, orbicular to orbicular, densely pubescent, apically concave; perianth persistent. Seed at center of samara.

Distribution in Beijing: Changping District: Huyu, Gou'ai; Mentougou District: Longmenjian Canyon, Changtiaoyu, Yanjiatai, Nanshiyang Canyon; Fangshan District: Shangfang Mountain, Puwa; Yanqing District: Xidazhuangke Village.

Protection Strategies: Protect the soil structure in its growth area, avoiding soil compaction and root rot caused by excessive moisture. Establish protected zones within its distribution area to minimize human disturbance. Conduct research on artificial propagation to scientifically improve seed germination rates and seedling survival rates, gradually expanding the population size.

032

中文名 **青檀**

学　名 *Pteroceltis tatarinowii*

大麻科
Cannabaceae

青檀属
Pteroceltis

[形态描述] 乔木，高达20米；树皮灰色或深灰色；皮孔明显；冬芽卵形。叶宽卵形至长卵形，先端渐尖至尾状渐尖，基部不对称，边缘有不整齐的锯齿，叶面绿，幼时被短硬毛，后脱落，常残留有圆点。翅果状坚果近圆形或近四方形，黄绿色或黄褐色，果实外面无毛或多少被曲柔毛，具宿存的花柱和花被。花期3～5月，果期8～10月。

[北京分布] 海淀区（香山）；房山区（上方山，观音殿，至斗泉）；门头沟区（滴水岩，天泉寺）；昌平区（桃洼，夏口）；平谷区（金山）。

[保护策略] 保持生境的土壤结构和稳定性，防止土壤压实和水土流失；在其生长区域设立缓冲区或保护区域，尽量减少人为干扰；通过定期巡查，及早发现病虫害并采取生物防治或生态友好的病虫害管理措施。

Morphological Description: Trees, to 20 m tall. Bark gray or dark gray. Branchlets with distinct lenticels. Winter buds ovoid. Leaf blade broadly ovate to oblong, base oblique, margin irregularly serrate, apex acuminate. Nut yellowish green to yellowish brown, globose to oblong, glabrous or pubescent; perianth and style persistent. Fl. Mar–May, fr. Aug–Oct.

Distribution in Beijing: Haidian District: Xiangshan; Fangshan District: Shangfang Mountain, Guanyindian, Zhidouquan; Mentougou District: Dishuiyan, Tianquan Temple; Changping District: Taowa, Xiakou; Pinggu District: Jinshan Mountain.

Protection Strategies: Maintain the soil structure and stability of the habitat, preventing soil compaction and erosion. Establish buffer zones or protected areas in its growth region to minimize human disturbance. Conduct regular inspections to detect pests and diseases early and implement biological control or environmentally friendly pest management measures.

033

中文名 **柘**（柘树）

学　名 *Maclura tricuspidata*

桑科
Moraceae

橙桑属
Maclura

[形态描述] 落叶灌木或小乔木，高1～7米；树皮灰褐色，小枝无毛，略具棱；冬芽赤褐色。叶卵形或菱状卵形，先端渐尖，基部楔形至圆形，无毛或被柔毛。雌雄异株，雌雄花序均为球形头状花序，单生或成对腋生；雄花花被片肉质，先端肥厚，内卷，退化雌蕊锥形；花被片先端盾形，内卷，子房埋于花被片下部。聚花果近球形，肉质，成熟时橘红色。花期5～6月，果期6～7月。

[北京分布] 房山区（张坊镇）；门头沟区（潭柘寺）；大兴区（黄村镇）；海淀区（罗道庄）。

[保护策略] 在其分布区加强水土保持，避免土地开发导致的土壤流失和干旱；通过增施有机肥保持土壤肥力；在其自然分布区设立保护区或缓冲区；进行定期监测，发现病虫害时优先放置捕虫器或采用生物防治的方法。

Morphological Description: Shrubs or small trees, 1–7 m tall. Bark grayish brown. Branchlets slightly ridged, glabrous. Winter buds reddish brown. Leaf blade ovate to rhombic-ovate, abaxially greenish white and glabrous or sparsely pubescent, base rounded to cuneate, apex acuminate. Inflorescences axillary, single or in pairs. Male inflorescences capitulate. Female inflorescence; peduncle short. Male flowers: calyx lobes fleshy, margin revolute, apex thick; pistillode pyramidal. Female flowers: calyx lobes with margin revolute, apically shield-shaped; ovary immersed in lower part of calyx. Fruiting syncarp orange red when mature, globose. Fl. May–Jun, fr. Jun–Jul.

Distribution in Beijing: Fangshan District: Zhangfang Town; Mentougou District: Tanzhe Temple; Daxing District: Huangcun Town; Haidian District: Luodaozhuang.

Protection Strategies: Strengthen soil and water conservation in its distribution area to prevent soil erosion and drought caused by land development. Maintain soil fertility by applying organic fertilizers. Establish protected areas or buffer zones within its natural distribution range. Conduct regular monitoring, and prioritize the use of insect traps or biological control methods when pests and diseases are detected.

034

中文名 **千金榆**

学　名 *Carpinus cordata*

桦木科
Betulaceae

鹅耳枥属
Carpinus

[形态描述]　乔木，高约18米；树皮灰色；小枝棕色或橘黄色，具沟槽。叶厚纸质、卵形或矩圆状卵形，顶端渐尖，基部斜心形，边缘具不规则的刺毛状重锯齿，上面疏被长柔毛或无毛，下面沿脉疏被短柔毛，侧脉15～20对。果序轴密被短柔毛及稀疏的长柔毛；果苞宽卵状矩圆形，其边缘的上部具疏齿，内侧的边缘具明显的锯齿。小坚果矩圆形，无毛，具不明显的细肋。

[北京分布]　密云区（坡头，卧龙山，曹家路村）；平谷区（镇罗营镇）。

[保护策略]　保护其分布区域的土壤环境，避免大规模开发和植被破坏造成土壤退化，必要时进行土壤改良，调节酸碱度；千金榆易受到病虫害威胁，宜采用生物防治方法或设置物理捕虫装置。

Morphological Description: Trees to 18 m tall; bark gray or black-gray, scaly fissured. Branchlets brown or yellow-brown. Leaf blade ovate, ovate-oblong, or obovate-oblong, abaxially sparsely to densely villous along midvein and lateral vein, adaxially sparsely villous or glabrescent, base unequally cordate, margin irregularly, apex acuminat; lateral veins 15–20 on each side of midvein. Female inflorescence sparsely pubescent or glabrescent; bracts broadly ovate-oblong, outer margin remotely serrate and inflexed, inner margin remotely serrate distally. Nutlet oblong, glabrous, faintly ribbed.

Distribution in Beijing: Miyun District: Potou, Wolong Mountain, Caojialu Village; Pinggu District: Zhenluoying Town.

Protection Strategies: Protect the soil environment in its distribution area, avoiding soil degradation caused by large-scale development and vegetation destruction. Improve the soil when necessary, regulate pH levels. *Carpinus cordata* is susceptible to pests and diseases, and biological control methods or physical trapping devices are recommended.

035

中文名 **铁木**
学　名 *Ostrya japonica*

桦木科
Betulaceae

铁木属
Ostrya

[形态描述] 乔木，高达20米。树皮暗灰色；枝条暗灰褐色，皮孔疏生；密被短柔毛。叶卵形至卵状披针形，顶端渐尖，基部近圆形、心形、斜心形或宽楔形；边缘具不规则的重锯齿；上面绿色，疏被短柔毛或几无毛。雄花序单生叶腋间或2~4枚聚生；苞鳞宽卵形。小坚果长卵圆形，淡褐色，有光泽，具数肋，无毛。

[北京分布] 怀柔区（喇叭沟门）；密云区（五座楼）。

[保护策略] 在其生长区加强水土保持，在干旱季节通过适度灌溉保持土壤湿润，并增加土壤有机质含量以保持肥力；定期巡查铁木的健康状况，尽早发现病虫害，采取生物防治方法。

Morphological Description: Trees, up to 20 m tall. Bark dark gray; branches dark gray-brown, with sparse lenticels; densely covered with short pubescence. Leaves ovate to ovate-lanceolate, apex acuminate, base subrounded, cordate, obliquely cordate, or broadly cuneate; margin with irregular double serrations; upper surface green, sparsely pubescent or subglabrous. Male inflorescences solitary in leaf axils or clustered in 2–4; bract scales broad ovate. Nutlets long ovoid, light brown, glossy, with several ribs, glabrous.

Distribution in Beijing: Huairou District: Labagoumen; Miyun District: Wuzuolou.

Protection Strategies: Strengthen soil and water conservation in its growing areas, maintain soil moisture during dry seasons through moderate irrigation, increase soil organic matter content to preserve fertility. Regularly monitor the health status of *Ostrya japonica* to detect pests and diseases early, and adopt biological control methods.

036 中文名 **梧桐杨**
学　名 *Populus pseudomaximowiczii*

杨柳科
Salicaceae

杨属
Populus

[形态描述]　乔木，高15米。树皮灰色，附白霜。小枝粗壮，赤褐色或黄赤褐色。萌枝叶宽卵形或卵状椭圆形，先端突尖，基部心形，边缘有不整齐粗腺齿缘，上面暗绿色，下面苍白；短枝叶阔卵形或卵形，先端突尖或短渐尖，常扭曲，基部浅心形或近圆形，边缘圆锯齿，有缘毛。雄花序轴具毛；苞片褐色，丝裂，无毛。蒴果卵圆形，被柔毛，3（2）瓣裂近无柄。花期4月，果期6月。

[北京分布]　密云区（雾灵山）。

[保护策略]　在其生长区采取水土保持措施，必要时进行适当灌溉，维持土壤湿度；在其自然生长区设立保护区域或限制采伐；对梧桐杨的生长区开展长期监测，根据监测数据调整保护措施；定期检查梧桐杨的健康状况。

Morphological Description: Trees to 15 m tall; bark gray, pruinose. Branchlets russet or yellowish russet, stout. Leaves of short branchlets blade broadly ovate or ovate, abaxially pale, adaxially dark green, both surfaces whitish long pubescent along veins, base shallowly cordate or subrounded, margin crenate, ciliate, apex apiculate, abaxially pale, adaxially dark green, base cordate, margin irregularly and coarsely glandular dentate, apex apiculate. Male inflorescence axis is pubescent; bracts brown, laciniate, glabrous. Capsule ovoid, downy, rarely glabrous, (2 or)3-valved. Fl. Apr, fr. Jun.

Distribution in Beijing: Miyun District: Wuling Mountain.

Protection Strategies: Implement soil and water conservation measures in its growing areas, providing appropriate irrigation when necessary to maintain soil moisture. Establish protected areas or restrict logging in its natural habitats. Conduct long-term monitoring of *Populus pseudomaximowiczii* growth areas and adjust conservation measures based on monitoring data. Regularly inspect the health status of *Populus pseudomaximowiczii*.

037

中 文 名 **省沽油**
学　　名 *Staphylea bumalda*

省沽油科
Staphyleaceae

省沽油属
Staphylea

[形态描述] 落叶灌木，高约2米，树皮紫红色或灰褐色，有纵棱；枝条开展，绿白色复叶对生，具三小叶；小叶椭圆形、卵圆形或卵状披针形，具尖尾，基部楔形或圆形，边缘有细锯齿，上面无毛，背面青白色。圆锥花序顶生，直立；萼片长椭圆形，浅黄白色；花瓣5，白色，倒卵状长圆形，较萼片稍大；雄蕊5，与花瓣略等长。种子黄色，有光泽。花期4~5月，果期8~9月。

[北京分布] 房山区（上方山，地藏庵）；昌平区（虎峪，九仙庙）；平谷区（金山）。

[保护策略] 在其分布区域控制土壤湿度；增加土壤有机肥含量，保持良好的肥力和透气性；在其分布区域设立保护区或缓冲区；定期检查省沽油的健康状况；开展长期的种群监测。

Morphological Description: Shrubs, 2 m tall; bark dark red or grayish brown with vertical stripes. Branches spreading, greenish white. Leaves trifoliolate; leaflet blades elliptic, ovate, or lanceolate-ovate, glabrescent and green on adaxial surface, pubescent along veins and pale green on abaxial surface, base cuneate or rounded, margin serrulate with sharp teeth. Inflorescence a terminal panicle, erect. Sepals light yellow, elliptic. Petals 5, white, obovate, slightly larger than sepals. Stamens 5, as long as petals. Ovary 2-locular. Seeds shining yellow. Fl. Apr–May, fr. Aug–Sep.

Distribution in Beijing: Fangshan District: Shangfang Mountain, Dizang Temple; Changping District: Huyu, Jiuxian Temple; Pinggu District: Jinshan Mountain.

Protection Strategies: Control soil moisture in its distribution area. Increase the organic fertilizer content in the soil to maintain good fertility and aeration. Establish protected areas or buffer zones in its distribution range. Regularly inspect the health status of *Staphylea bumalda*. Conduct long-term population monitoring.

038

中文名 **黄连木**
学　名 *Pistacia chinensis*

漆树科
Anacardiaceae

黄连木属
Pistacia

[形态描述] 落叶乔木，高达20余米；树皮暗褐色，幼枝疏被微柔毛。奇数羽状复叶互生，小叶对生或近对生，纸质，披针形，先端渐尖或长渐尖，基部偏斜，全缘。先花后叶，雄花序排列紧密，雌花序排列疏松，苞片披针形或狭披针形；雄花花被片2~4，披针形或线状披针形；雌花花被片7~9，披针形或线状披针形，里面5片卵形或长圆形。核果倒卵状球形，成熟时紫红色。

[北京分布] 海淀区（鹫峰，八大处）；房山区（上方山，圣水峪，地藏庵）；石景山区（四平台）。

[保护策略] 改善土壤透气性，并保持适宜的湿度；避免不当采摘和踩踏；定期对黄连木进行病虫害检查；对其种群健康状况和环境数据进行长期监测，并结合气候变化趋势，制定应对措施。

Morphological Description: Deciduous trees, up to over 20 m tall; bark dark brown, young branches sparsely pubescent with minute hairs. Odd-pinnately compound leaves alternate, leaflets opposite or subopposite, papery, lanceolate, apex acuminate or long acuminate, base oblique, margin entire. Flowers produced before leafing; male inflorescences densely arranged, female inflorescences laxly arranged; bracts lanceolate or narrowly lanceolate; Male flowers with 2–4 tepals, lanceolate or linear-lanceolate; Female flowers with 7–9 tepals, lanceolate or linear-lanceolate, inner 5 tepals ovate or oblong. Drupe obovate-spherical, purplish red when ripe.

Distribution in Beijing: Haidian District: Jiufeng, Badachu; Fangshan District: Shangfang Mountain, Shengshuiyu, Dizang Temple; Shijingshan District: Sipingtai.

Protection Strategies: Improve soil aeration and maintain appropriate moisture levels. Avoid improper harvesting and trampling. Regularly inspect *Pistacia chinensis* for pests and diseases. Conduct long-term monitoring of its population health and environmental data, and develop response measures based on climate change trends.

039 中文名 **漆**（漆树）

学　名 *Toxicodendron vernicifluum*

漆树科
Anacardiaceae

漆树属
Toxicodendron

[形态描述] 落叶乔木，高达20米。奇数羽状复叶互生；小叶卵形、卵状椭圆形或长圆形，先端急尖或渐尖，全缘，叶面通常无毛。圆锥花序被灰黄色微柔毛；花萼无毛，裂片卵形；花瓣长圆形，具细密的褐色羽状脉纹，开花时外卷；雄蕊花丝线形；子房球形，花柱3。果序多少下垂，核果肾形或椭圆形，先端锐尖，基部截形，外果皮成熟后不裂。花期5~6月，果期7~10月。

[北京分布] 昌平区（南口镇沙岭，下口村，大杨山森林公园，兴寿镇）；房山区（上方山）。

[保护策略] 改善其生境内的土壤透气性，防止水分积聚；定期检查漆树的健康状况，发现病虫害及时防治；在漆树生长区设立保护小区，以避免人为采割树脂对树体造成影响。

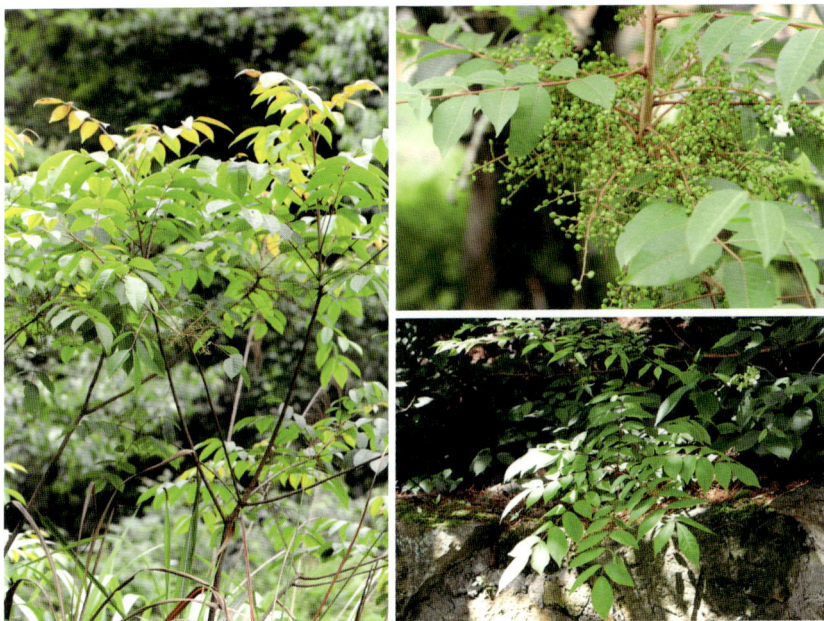

Morphological Description: Deciduous trees to 20 m tall; leaf blade imparipinnately compound; leaflet blade ovate to ovate-elliptic or oblong, adaxially glabrous, base oblique, rounded or broadly cuneate, margin entire, apex acute or acuminate. Inflorescence paniculate, grayish yellow minutely pubescent. Calyx lobes ovate. Petals yellowish green, oblong, with brown featherlike venation pattern. Stamens filaments equal to anthers in length. Ovary globose; styles 3. Infructescence pendulous; drupe kidney-shaped or elliptical, sharply pointed at the apex and truncate at the base. Epicarp indehiscent when ripe. Fl. May–Jun, fr. Jul–Oct.

Distribution in Beijing: Changping District: Shaling in Nankou Town, Xiakou Village, Dayangshan Forest Park, Xingshou Town; Fangshan District: Shangfang Mountain.

Protection Strategies: Improve soil aeration in its habitat to prevent water accumulation. Regularly inspect the health of *Toxicodendron vernicifluum*, and promptly control pest and disease upon discovery. Establish protection plots in the *Toxicodendron vernicifluum* growth regions of Beijing to avoid the impact of human resin harvesting on the trees.

040 中文名 **葛萝槭**
　　学　名 *Acer davidii*

无患子科
Sapindaceae

槭属
Acer

[形态描述] 落叶乔木。叶纸质，卵形，边缘具密而尖锐的重锯齿，基部近于心脏形，5裂。花淡黄绿色，单性，雌雄异株，常呈细瘦下垂的总状花序；萼片5，花瓣5，雄蕊8，子房紫色。翅果嫩时淡紫色，成熟后黄褐色；小坚果略微扁平；翅张开呈钝角或近于水平。

[北京分布] 怀柔区（龙潭沟，宝山寺）；延庆区（珍珠泉乡三岔口村）；密云区（坡头林场）；房山区（上方山）。

[保护策略] 在其分布区域保持适宜的土壤湿度；在必要时添加腐殖质，增加土壤肥力；及时发现和防治病虫害；长期监测并定期评估其健康状态。

Morphological Description: Deciduous trees. Leaves papery, ovate, with dense and sharp double serrations on the margin, base nearly cordate, and 5-lobed. Flowers pale yellow-green, unisexual, dioecious, often in slender, pendulous racemes; sepals 5, petals 5, stamens 8, ovary purple. Samaras light purple when young, yellow-brown when mature; nutlet slightly flattened; wings spread at an obtuse angle or nearly horizontal.

Distribution in Beijing: Huairou District: Longtangou, Baoshan Temple; Yanqing District: Sanchakou Village, Zhenzhuquan Township; Miyun District: Potou Forest Farm; Fangshan District: Shangfang Mountain.

Protection Strategies: Maintain appropriate soil moisture in its distribution areas. Add humus when necessary to enhance soil fertility. Detect and control pests and diseases promptly. Conduct long-term monitoring and regularly assess its health status.

041

中文名 **青花椒**

学　名 *Zanthoxylum schinifolium*

芸香科
Rutaceae

花椒属
Zanthoxylum

[形态描述] 通常为高 1～2 米的灌木；茎枝有短刺，嫩枝暗紫红色。叶有小叶 7～19 片；小叶对生，宽卵形至披针形。花序顶生；萼片及花瓣均 5 片；花瓣淡黄白色；雄花的退化雌蕊甚短。雌花有心皮 3 个，很少 4 或 5 个。分果瓣红褐色，干后变暗苍绿或褐黑色，顶端几无芒尖，油点小。花期 7～9 月，果期 9～12 月。

[北京分布] 房山区（黄山店村）；平谷区（熊耳寨，南独乐河镇，石林峡，白羊村）；密云区（云蒙峡）。

[保护策略] 在土壤中添加有机质和沙质材料，增强排水性，并适度改良土壤结构；定期剪去受到蚜虫和根腐病等病虫害影响的枝叶，清理落叶；在冬季低温期间，覆盖植株基部以防冻害，在高温季节提供遮阳设施或适量灌溉。

Morphological Description: Shrubs 1–2 m tall. Stems and branchlets with prickles. Young branchlets dark purplish red. Leaves 7–19-foliolate; leaflet blades opposite, broadly ovate, broadly ovate-rhombic, or lanceolate. Inflorescences terminal. Flowers 5-merous. Petals pale yellowish white. Male flowers: rudimentary gynoecium. Female flowers: carpels 3(–5). Fruit follicles reddish brown but dark green to brownish black when dry, oil glands small, apex not beaked. Fl. Jul–Sep, fr. Sep–Dec.

Distribution in Beijing: Fangshan District: Huangshandian Village; Pinggu District: Xiong'erzhai, Nandulehe Town, Shilin Gorge, Baiyang Village; Miyun District: Yunmeng Gorge.

Protection Strategies: Add organic matter and sandy materials to the soil to enhance drainage and moderately improve soil structure. Regularly prune branches and leaves affected by pests such as aphids or diseases like root rot, and clear fallen leaves. During low winter temperatures, cover the base of the plants to prevent frost damage, and in hot seasons, provide shading or moderate irrigation.

129

042 中文名 **白鲜**
学　名 *Dictamnus dasycarpus*

芸香科
Rutaceae

白鲜属
Dictamnus

[形态描述] 茎基部木质化的多年生宿根草本，高40～100厘米。叶有小叶9～13片，小叶对生，无柄，椭圆至长圆形，叶缘有细锯齿，叶脉不甚明显。总状花序；苞片狭披针形；花瓣白带淡紫红色或粉红带深紫红色脉纹，倒披针形；雄蕊伸出于花瓣外；萼片及花瓣均密生透明油点。成熟的果（蓇葖）沿腹缝线开裂为5个分果瓣；种子阔卵形或近圆球形。花期5月，果期8～9月。

[北京分布] 延庆区（海坨山）。

[保护策略] 在其种植区进行土壤改良，通过添加腐殖质或有机肥料提升土壤肥力和水分保持能力；在白鲜的生长季节加强管理，避免因不当采挖导致资源损耗；在多雨季节保持土壤排水通畅，避免积水。

Morphological Description: Perennial herbaceous plants with woody basal stems, perennial rootstock, 40–100 cm tall. Leaves with 9–13 leaflets; leaflets opposite, sessile, elliptic to oblong, margin with fine serrations, leaf veins not very distinct. Racemose inflorescence; bracts narrowly lanceolate; petals white with pale purplish-red or pink with deep purplish-red vein patterns, oblanceolate; stamens exserted beyond petals; both sepals and petals densely covered with transparent oil dots. Mature fruit (follicles) dehisce along ventral sutures into 5 mericarps; seeds broadly ovate or subglobose. Fl. May, fr. Aug–Sep.

Distribution in Beijing: Yanqing District: Haituo Mountain.

Protection Strategies: Improve the soil in its growing areas by adding humus or organic fertilizers to enhance soil fertility and water retention. Strengthen management during the growing season of *Dictamnus dasycarpus* to prevent resource depletion caused by improper harvesting. Ensure proper drainage during rainy seasons to avoid waterlogging.

043

中文名 **宽苞水柏枝**

学 名 *Myricaria bracteata*

柽柳科
Tamaricaceae

水柏枝属
Myricaria

[形态描述] 灌木，高0.5～3米，多分枝；老枝灰褐色或紫褐色。叶卵形、卵状披针形、线状披针形或狭长圆形，先端钝或锐尖。总状花序顶生；苞片通常宽卵形或椭圆形，先端渐尖，基部狭缩；萼片披针形，长圆形或狭椭圆形，先端钝或锐尖；花瓣倒卵形或倒卵状长圆形，粉红色、淡红色或淡紫色；雄蕊略短于花瓣；子房圆锥形。蒴果狭圆锥形。种子狭长圆形或狭倒卵形。花期6～7月，果期8～9月。

[北京分布] 延庆区（玉渡山）；门头沟区（东灵山）。

[保护策略] 迁地保护时宜选择排水性能好的斜坡或略高的地形，避免积水，并增强土壤的通气性和保水性；定期检查植株健康状况，减少叶面湿度，预防病菌传播；在寒冷季节为植株基部增加覆盖物，以防受冻害。

Morphological Description: Shrubs, 0.5–3 m tall, much branched. Old branches gray-brown or purple-brown. Leave ovate, ovate-lanceolate, linear-lanceolate, or narrowly oblong, apex obtuse or acute. Racemes terminal on branches of current year; bracts usually broadly ovate or elliptic, base narrow, apex acuminate. Sepals lanceolate, oblong, or narrowly elliptic, margin broadly membranous. Petals pink, reddish, or purplish, obovate or obovate-oblong. Stamens slightly shorter than petals. Ovary conic. Capsule narrowly conic. Seeds narrowly oblong or narrowly obovate. Fl. Jun–Jul, fr. Aug–Sep.

Distribution in Beijing: Yanqing District: Yudu Mountain; Mentougou District: Dongling Mountain.

Protection Strategies: In ex-situ conservation, it is advisable to select slopes or slightly elevated terrain with good drainage to avoid waterlogging, and enhance soil aeration and water retention. Regularly check plant health, reduce leaf surface humidity to prevent the spread of pathogens. During the cold season, add mulch around the base of plants to help them withstand severe cold.

044

中文名 **八角枫**
学　名 *Alangium chinense*

山茱萸科
Cornaceae

八角枫属
Alangium

[形态描述]　落叶乔木或灌木，高3～5米；幼枝无毛或有稀疏的疏柔毛。叶近圆形或椭圆形、卵形，顶端短锐尖，基部两侧常不对称，阔楔形、截形，稀近于心形；基出脉3～5（～7），侧脉3～5对。聚伞花序腋生；花萼顶端分裂为5～8枚齿状萼片；花瓣6～8，线形；雄蕊和花瓣同数。核果卵圆形，种子1颗。花期5～7月和9～10月，果期7～11月。

[北京分布]　海淀区（卧佛寺）。

[保护策略]　在其栽培区域应使用富含有机质的土壤，增加土壤的保水性和营养供应；选择排水性好的坡地或稍高地势，避免积水和土壤板结；定期检查植株，采用生物防治方法治理病虫害；使用覆盖物为植株根部和基部保温，减少霜冻伤害。

Morphological Description: Deciduous trees or shrubs, 3–5 m tall; branchlets glabrous or sparsely pubescent when young. Leaves suborbicular, elliptic, or ovate, apex shortly acute, base often asymmetric, broadly cuneate, truncate, rarely subcordate; basal veins 3–5 (–7), lateral veins 3–5 pairs. Cymes axillary; calyx apex divided into 5–8 dentate lobes; petals 6–8, linear; stamens equal in number to petals. Drupe ovoid, seed 1. Fl. May–Jul and Sep–Oct, fr. Jul–Nov.

Distribution in Beijing: Haidian District: Wofo Temple.

Protection Strategies: In its cultivation areas, organic-rich soil should be used to enhance water retention and nutrient supply. Select well-drained slopes or slightly elevated terrain to avoid waterlogging and soil compaction. Regularly inspect the plants and apply biological control methods to manage pests and diseases. Use mulch to insulate the roots and base of the plants, reducing frost damage.

045

中 文 名 **箭报春**
学　　名 *Primula fistulosa*

报春花科
Primulaceae

报春花属
Primula

[**形态描述**] 多年生草本。叶片矩圆形至矩圆状倒披针形，先端渐尖或稍钝，基部渐狭窄，边缘具不整齐的浅齿。花葶中空，顶部（花序下）缢缩；伞形花序通常多花，密集呈球状；苞片多数，矩圆状卵形或卵状披针形，先端多少锐尖；花萼钟状或杯状，自基部向上渐增宽，裂片矩圆状披针形，先端锐尖；花冠玫瑰红色或红紫色；长花柱花：雄蕊着生于冠筒中部，花柱长达冠筒口；短花柱花：雄蕊着生于冠筒中上部。蒴果球形，与花萼近等长。花期5～6月。

[**北京分布**] 延庆区（玉渡山，松山）。

[**保护策略**] 在栽培区域加入腐殖质和有机肥，增强土壤的保水能力和营养供应；在植株周围铺设有机覆盖物，可减少水分蒸发，且能改善土壤湿度；通过标识和适当围栏限制采摘行为；定期检查植株健康状况。

Morphological Description: Herbs perennial; leaf blade oblong to oblong-oblanceolate, base attenuate, margin irregularly denticulate, apex acuminate to acute. Scapes fistular, constricted below inflorescences; umbels globose, usually many flowered; bracts many, oblong-ovate to ovate-lanceolate, apex acute. Calyx cup-shaped, gradually dilated upward from base; lobes oblong-lanceolate, apex acute. Corolla rose-violet. Pin flowers: stamens at middle of corolla tube; style reaching mouth. Thrum flowers: stamens in upper 1/2 of corolla tube. Capsule globose, nearly as long as calyx. Fl. May–Jun.

Distribution in Beijing: Yanqing District: Yudu Mountain, Songshan Mountain.

Protection Strategies: In the cultivation area, incorporate humus and organic fertilizers to enhance the soil's water retention capacity and nutrient supply. Apply organic mulch around the plants to reduce water evaporation and improve soil moisture. Limit harvesting activities through signage and appropriate fencing. Regularly inspect the health of the plants.

046

中文名 粉报春
学　名 *Primula farinosa*

报春花科
Primulaceae

报春花属
Primula

[形态描述] 多年生草本。叶多数，形成较密的莲座丛，叶片矩圆状倒卵形、窄椭圆形或矩圆状披针形，先端近圆形或钝，基部渐狭窄，边缘具稀疏小牙齿或近全缘，近顶端通常被青白色粉。伞形花序顶生，通常多花；苞片狭披针形或先端渐尖成钻形，基部增宽并稍膨大；花萼钟状，内面通常被粉，边缘具短腺毛；花冠淡紫红色，裂片楔状倒卵形，先端2深裂；长花柱花；雄蕊着生于冠筒中部；短花柱花；雄蕊着生于冠筒中上部。蒴果筒状。

[北京分布] 延庆区（滴水壶）。

[保护策略] 保护生境，减少人为干扰和采摘；注意人工采种繁育，适时回归。

Morphological Description: Herbs perennial. Leaves numerous, forming a dense rosette; leaf blade oblong-obovate to oblong-lanceolate, base attenuate, margin remotely denticulate to nearly entire, apex subrounded to obtuse. Usually farinose toward apex; umbels usually many flowered; bracts narrowly lanceolate to acuminate-subulate, base dilated. Calyx campanulate, usually farinose inside. Corolla lilac to purple; lobes cuneate-obovate, deeply emarginate. Pin flowers: stamens at middle of corolla tube. Thrum flowers: stamens in upper 1/2 of corolla tube.

Distribution in Beijing: Yanqing District: Dishuihu.

Protection Strategies: Protect habitats by reducing human disturbance and picking. Implement ex situ conservation through artificial seed collection, followed by strategic reintroduction.

047 中文名 **岩生报春**
学　名 *Primula saxatilis*

报春花科
Primulaceae

报春花属
Primula

[形态描述] 多年生草本，叶3～8枚丛生，叶片阔卵形至矩圆状卵形，先端钝，基部心形，边缘具缺刻状或羽状浅裂，裂片边缘有三角形牙齿。伞形花序1～2轮，每轮3～9（15）花；苞片线形至矩圆状披针形，疏被短柔毛；花梗被柔毛或短柔毛；花萼近筒状，疏被短毛，分裂达中部，裂片直立，具明显的中肋；花冠淡紫红色，裂片倒卵形，先端具深凹缺；长花柱花：雄蕊着生于冠筒中下部，花柱长略低于冠筒口；短花柱花：雄蕊稍低于喉部环状附属物，花柱长达冠筒中部。花期5～6月。

[北京分布] 密云区（遥桥峪，云岫谷，坡头）。

[保护策略] 保护生境，减少人为干扰和采摘；注意人工采种繁育，适时回归。

Morphological Description: Herbs perennial. Leaves 3–8 tufted; leaf blade broadly ovate to oblong-ovate, base cordate, margin incised or pinnately lobulate, apex obtuse; lobules with triangular teeth. Umbels 1 or 2, 3–9(–15)-flowered; bracts linear to oblong-lanceolate, sparsely pubescent. Pedicel pilose or pubescent. Calyx tubular-campanulate, sparsely pubescent or glabrous, parted to middle; lobes erect, conspicuously costate. Corolla rose-purple; lobes obovate, deeply emarginate. Pin flowers: stamens near middle of corolla tube; style exserted beyond annulus. Thrum flowers: stamens slightly below annulus; style reaching middle of corolla tube. Fl. May–Jun.

Distribution in Beijing: Miyun District: Yaoqiaoyu, Yunxiugu, Potou.

Protection Strategies: Protect habitats by reducing human disturbance and picking. Implement ex situ conservation through artificial seed collection, followed by strategic reintroduction.

048

中文名 **葛枣猕猴桃**
学　名 *Actinidia polygama*

猕猴桃科
Actinidiaceae

猕猴桃属
Actinidia

[**形态描述**] 大型落叶藤本；着花小枝基本无毛；髓白色，实心。叶卵形或椭圆卵形，顶端急渐尖至渐尖，基部圆形或阔楔形，边缘有细锯齿，腹面绿色，背面浅绿色；叶柄近无毛。花序1～3花；花白色；萼片5片，卵形至长方卵形，两面薄被微茸毛或近无毛；花瓣5片，倒卵形至长方倒卵形；花药黄色，卵形；子房瓶状。果成熟时淡橘色，卵珠形或柱状卵珠形，顶端有喙。花期6月中旬至7月上旬，果熟期9～10月。

[**北京分布**] 房山区（上方山）；昌平区（十三陵）。

[**保护策略**] 保护生境，减少人为干扰和采摘；注意人工采种繁育，适时回归。

Morphological Description: Climbing shrubs, large, deciduous. Branchlets glabrous; pith white, large, solid. Petiole glabrous; leaf blade abaxially pale green, adaxially green to entirely white, ovate to oblong-ovate, base broadly cuneate to rounded, margin serrulate, apex acuminate to abruptly acuminate. Inflorescences 1–3-flowered in a fascicle. Flowers white. Sepals 5, ovate to oblong-ovate, both surfaces glabrous or sparsely puberulent. Petals 5, obovate to oblong-obovate; anthers yellow, ovoid. Ovary bottle-shaped. Fruit orange when mature, ovoid to cylindric-ovoid to oblong-ovoid, rostrate at apex. Fl. Jun–Jul, fr. Sep–Oct.

Distribution in Beijing: Fangshan District: Shangfang Mountain; Changping District: Ming Tombs.

Protection Strategies: Protect habitats by reducing human disturbance and picking. Implement ex situ conservation through artificial seed collection, followed by strategic reintroduction.

049

中文名 **鹿蹄草**
学　名 *Pyrola calliantha*

杜鹃花科
Ericaceae

鹿蹄草属
Pyrola

[形态描述] 常绿草本状小半灌木；根茎细长，横生，有分枝。叶4~7，基生，革质；椭圆形或圆卵形，先端钝头，基部阔楔形或近圆形，边缘近全缘或有疏齿，上面绿色，下面常有白霜。花葶有1~2（~4）枚鳞片状叶，卵状披针形或披针形。总状花序密生，花稍下垂，白色，有时稍带淡红色；花梗腋间有长舌形苞片；萼片舌形，边缘近全缘；花瓣倒卵状椭圆形或倒卵形；花柱常带淡红色。蒴果扁球形。花期6~8月；果期8~9月。

[北京分布] 密云区（坡头林场）；房山区（百花山）。

[保护策略] 保护生境，减少人为干扰和采摘；注意人工采种繁育，适时回归。

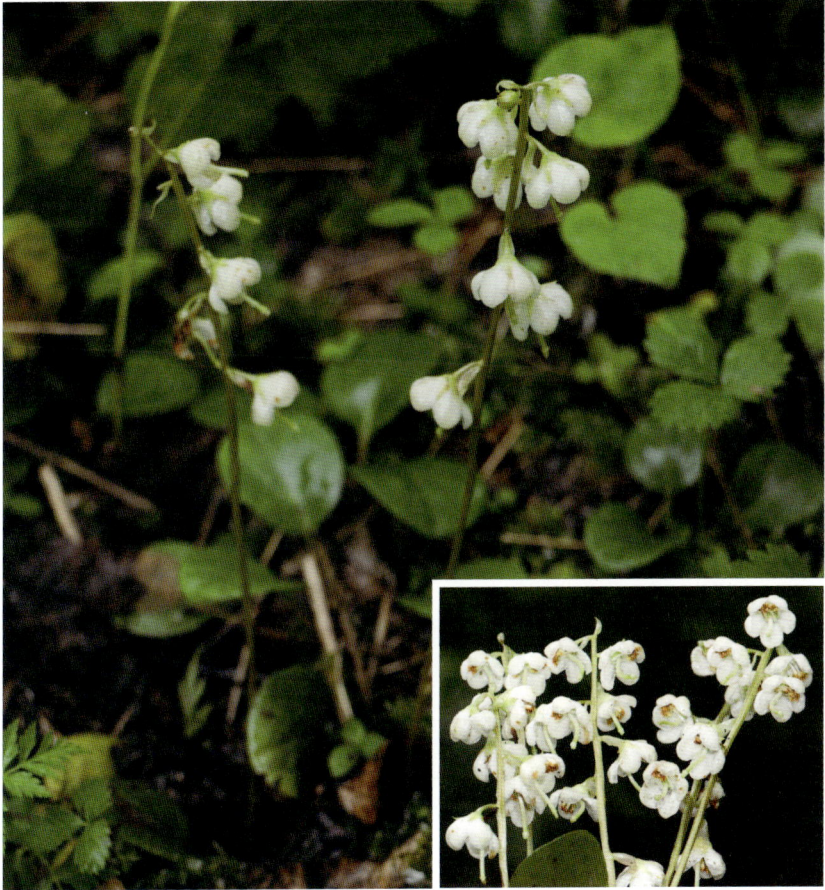

Morphological Description: Evergreen herb-like small suffrutices; rhizome long, slender, horizontal, branched. Leaves 4–7,basal, leathery; elliptic or orbicular-ovate, apex obtuse, base broadly cuneate or suborbicular, margin subentire or sparsely dentate, adaxially green, abaxially often glaucous. Scape with 1–2 (–4) scalelike leaves, ovate-lanceolate or lanceolate. Raceme dense, flowers slightly pendulous, white, sometimes slightly reddish; axils of pedicels with long ligulate bracts; sepals ligulate, margin subentire; petals obovate-elliptic or obovate; style often reddish. Capsule compressed globose. Fl. Jun–Aug, fr. Aug–Sep.

Distribution in Beijing: Miyun District: Poto Forest Farm; Fangshan District: Baihua Mountain.

Protection Strategies: Protect habitats by reducing human disturbance and picking. Implement ex situ conservation through artificial seed collection, followed by strategic reintroduction.

050 中文名 **红花鹿蹄草**
学　名 *Pyrola asarifolia* subsp. *incarnata*

杜鹃花科
Ericaceae

鹿蹄草属
Pyrola

[形态描述] 常绿草本状小半灌木；根茎细长，横生，有分枝。叶3–7，基生，近圆形、圆卵形或卵状椭圆形，先端圆钝，基部近圆形或圆楔形，边缘近全缘或有不明显的浅齿。总状花序有7～15花；花梗腋间有膜质苞片，披针形；萼片三角状宽披针形，先端渐尖；花瓣倒圆卵形；花柱倾斜，上部向上弯曲，顶端有环状突起。花期6～7月；果期8～9月。

[北京分布] 门头沟区（东灵山）。

[保护策略] 保护生境，减少人为干扰和采摘；注意人工采种繁育，适时回归。

Morphological Description: Evergreen herb-like small suffrutices; rhizome long, slender, horizontal, branched. Leaves 3–7, basal; suborbicular, orbicular-ovate, or ovate-elliptic, apex rounded to obtuse, base suborbicular or broadly cuneate, margin subentire or obscurely shallow-dentate. Raceme 7–15-flowered; axils of pedicels with membranous bracts, lanceolate. Sepals triangular broadly lanceolate, apex acuminate. Petals obovate. Style oblique, upper part curved upward, apex with a ring-like projection. Fl. Jun–Jul, fr. Aug–Sep.

Distribution in Beijing: Mentougou District: Dongling Mountain.

Protection Strategies: Protect habitats by reducing human disturbance and picking. Implement ex situ conservation through artificial seed collection, followed by strategic reintroduction.

051

中文名 **松下兰**

学　名 *Hypopitys monotropa*

杜鹃花科
Ericaceae

松下兰属
Hypopitys

[形态描述] 多年生腐生草本；根系繁密；全株高约20厘米，淡白色或淡黄色，肉质，干后变黑。叶鳞片状，近直立，上部的稀疏，向下部渐较紧密，卵状矩圆形或宽披针形，上部的往往有不整齐的锯齿。花3～8朵成初俯垂、后渐直立的总状花序，花冠筒状钟形；花瓣4～5，淡黄色，颇为肉质，基部囊状，顶端钝，有不整齐的锯齿，雄蕊8～10，短于花冠。蒴果椭圆状球形，直立向上。

[北京分布] 门头沟区（百花山）。

[保护策略] 定期巡查松下兰的自然栖息地，加强对野外生长区域的保护，严查非法采摘和市场非法买卖；限制开发和其他干扰活动。

Morphological Description: Perennial saprophytic herb; roots dense. The entire plant reaches about 20 cm in height, pale white or pale yellow, fleshy, turning black after drying. Leaves are scale like, nearly upright, with the upper leaves sparse and the lower leaves becoming progressively tighter; ovate-rounded or broadly lanceolate in shape, with the upper leaves often having irregular serrations. Flowers are 3–8 in number, forming an initially drooping, later becoming upright, racemose inflorescence, tubular bell-shaped; petals 4–5, pale yellow, quite fleshy, with a pouch-like base and a blunt apex with irregular serrations. Stamens 8–10, shorter than the corolla. The capsule is ellipsoid to globose, upright, and pointed upward.

Distribution in Beijing: Mentougou District: Baihua Mountain.

Protection Strategies: Regularly patrol the natural habitats of *Hypopitys monotropa*, strengthen the protection of wild growing areas, and strictly monitor illegal collection and the illegal trade in markets. Limit development and other disruptive activities.

052

中文名 **迎红杜鹃**

学　名 *Rhododendron mucronulatum*

杜鹃花科
Ericaceae

杜鹃花属
Rhododendron

[**形态描述**] 落叶灌木，高1～2米。幼枝细长，疏生鳞片。叶椭圆形或椭圆状披针形，顶端锐尖、渐尖或钝，边缘全缘或有细圆齿，基部楔形或钝，上面疏生鳞片。花序腋生枝顶或假顶生，1～3花，先叶开放，伞形着生；花萼5裂；花冠宽漏斗状，淡红紫色，外面被短柔毛；雄蕊10，稍短于花冠；子房5室，密被鳞片。蒴果长圆形。花期4～6月，果期5～7月。

[**北京分布**] 门头沟区（小龙门林场，妙峰山，东灵山）；房山区（大安山，百花山，史家营，白草畔）；密云区（雾灵山）；怀柔区（琉璃庙，喇叭沟门）；昌平区（长峪城）；延庆区（大青沟，松山）；平谷区（镇罗营）；海淀区（西山）。

[**保护策略**] 定期监测和清除迎红杜鹃栖息地内的外来入侵植物；定期巡查迎红杜鹃的生长区域，禁止非法采摘和贸易行为。

Morphological Description: Deciduous multi-branched shrubs, 1–2 m tall; young branches thin and long, sparsely scaly. Leaf blade thin, elliptic or elliptic-lanceolate; base cuneate or obtuse, entire or denticulate; apex acute, acuminate or obtuse; adaxial surface sparsely scaly. Inflorescences axillary, terminal or pseudoterminal, 1–3-flowered, opening before leaves, umbellate; calyx 5-lobed; corolla funnelform, pale reddish purple, outer surface pubescent; stamens 10, slightly shorter than corolla tube; ovary 5-locular, densely scaly. Capsule cylindric. Fl. Apr–Jun, fr. May–Jul.

Distribution in Beijing: Mentougou District: Xiaolongmen Forestry Farm, Miaofeng Mountain, Dongling Mountain; Fangshan District: Da'an Mountain, Baihua Mountain Shijiaying, Baicaopan; Miyun District: Wuling Mountain; Huairou District: Liulimiao, Labagoumen; Changping District: Changyucheng; Yanqing District: Daqinggou, Songshan Mountain; Pinggu District: Zhenluoying; Haidian District: Xishan.

Protection Strategies: Regularly monitor and remove invasive alien plants from the habitat of *Rhododendron mucronulatum*. Periodically patrol the growth area of *Rhododendron mucronulatum* to prohibit illegal harvesting and trade.

053

中文名 **秦艽（jiāo）**

学　名 *Gentiana macrophylla*

龙胆科
Gentianaceae

龙胆属
Gentiana

[形态描述] 多年生草本，高30～60厘米。枝少数丛生，直立或斜升。莲座丛叶卵状椭圆形或狭椭圆形，先端钝或急尖，基部渐狭，边缘平滑；茎生叶椭圆状披针形或狭椭圆形，先端钝或急尖，基部钝，边缘平滑。花多数，无花梗，簇生枝顶呈头状或腋生作轮状；花冠筒部黄绿色，冠檐蓝色或蓝紫色，裂片卵形或卵圆形，全缘。蒴果卵状椭圆形；种子红褐色，有光泽。花果期7～10月。

[北京分布] 门头沟区（东灵山，百花山，灰金坨，小龙门）；房山区（大滩，白草畔，霞云岭）；昌平区（老峪沟村）；延庆区（松山，闫家坪，珍珠泉乡，海坨山）；怀柔区（喇叭沟门塘泉沟，郑栅子）；密云区（遥桥峪，坡头）。

[保护策略] 保护栖息地周围的植被，保持一定的荫蔽度，避免强烈的阳光直射，确保土壤湿润；维持栖息地内的生物多样性，提高生态系统的稳定性。

Morphological Description: Perennial herbs, 30–60 cm tall. Stems few, tufted, erect or ascending. Basal leaves in rosette, ovate-elliptic or narrowly elliptic, apex obtuse or acute, base attenuate, margin smooth; cauline leaves elliptic-lanceolate or narrowly elliptic, apex obtuse or acute, base obtuse, margin smooth. Flowers numerous, sessile, clustered at stem apex in capitate or axillary whorls; corolla tube yellow-green, limb blue or blue-purple, lobes ovate or ovate-round, entire. Capsules ovate-elliptic; seeds reddish-brown, glossy. Fl. and fr. Jul–Oct.

Distribution in Beijing: Mentougou District: Dongling Mountain, Baihua Mountain, Huijintuo, Xiaolongmen; Fangshan District: Datan, Baicaopan, Xiayunling; Changping District: Laoyugou Village; Yanqing District: Songshan Mountain, Yanjiaping, Zhenzhuquan Township, Haituo Mountain; Huairou District: Tangquangou in Labagoumen area, Zhengzhazi; Miyun District: Yaoqiaoyu, Potou.

Protection Strategies: Protect the vegetation around the habitat, maintaining a certain level of shade to avoid direct exposure to strong sunlight, and ensure the soil remains moist. Maitain biodiversity within habitats, enhancing the stability of the ecosystem.

054

中文名 **紫花杯冠藤**

学　名 *Cynanchum purpureum*

夹竹桃科
Apocynaceae

鹅绒藤属
Cynanchum

[形态描述] 直立草本植物，略为分枝而互生，干后中空。叶对生，集生于分枝的顶端，线形或线状披针形，两面被疏长柔毛。聚伞花序伞状；花萼裂片披针形，基部内面有小腺体；花冠无毛，紫红色，裂片披针形；副花冠薄膜质，顶端有5个浅齿。蓇葖果长圆形，两端略狭。花期5~6月，果期6月。

[北京分布] 怀柔区（喇叭沟门，石洞子，龙潭沟，琉璃庙）；密云区（塘子村）。

[保护策略] 保护其生长环境，维持栖息地内物种多样性，增强生境的稳定性。

Morphological Description: Herbs erect. Stems few branched, hollow when dry. Leaves opposite, usually grouped at branch apex; leaf blade linear or linear-lanceolate, densely ciliate. Inflorescences terminal, umbel-like. Sepals lanceolate; small basal glands present. Corolla purple, glabrous; lobes linear-lanceolate. Corona pale, tubular, membranous. Follicles compressed fusiform, both ends acute. Seeds oblong. Fl. May–Jun, fr. Jun.

Distribution in Beijing: Huairou District: Labagoumen, Shidongzi, Longtangou, Liulimiao; Miyun District: Tangzi Village.

Protection Strategies: Protect their growth environment, maintain biodiversity within habitats, and enhance habitat stability.

055

中 文 名 **长筒滨紫草**
学　名 *Mertensia davurica*

紫草科
Boraginaceae

滨紫草属
Mertensia

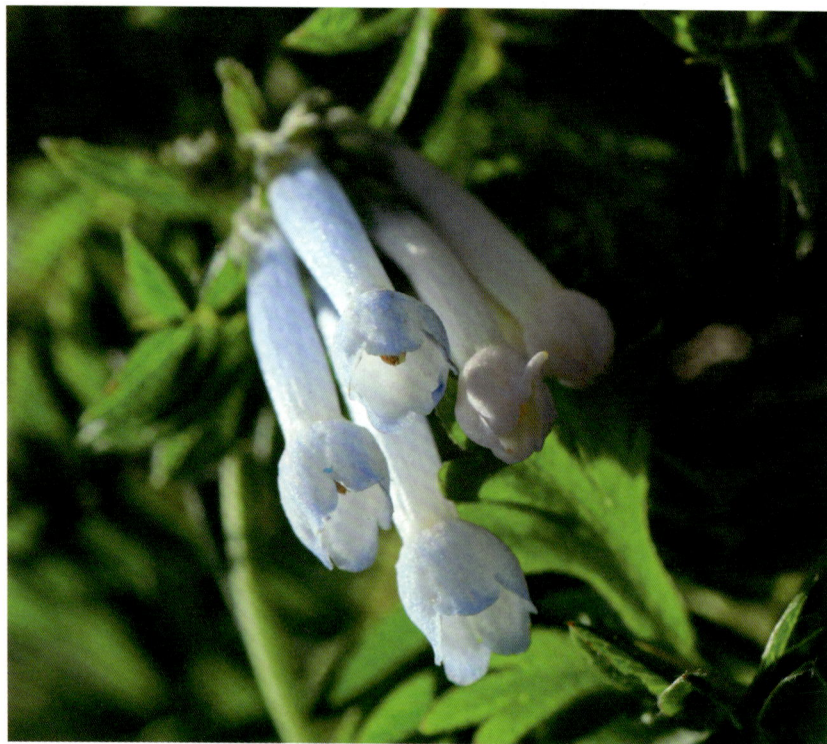

[**形态描述**] 多年生草本。根状茎块状，黑褐色。茎1条，直立。基生叶莲座状，叶片卵状长圆形或线状长圆形，基部楔形至圆形；茎生叶披针形至线状披针形，仅最下部的叶有柄而常早枯，先端钝或渐尖。镰状聚伞花序；花萼5裂至近基部，裂片线形或三角状线形；花冠蓝色，檐部比筒部稍宽，5浅裂，裂片近半圆形，全缘。小坚果有皱纹。

[**北京分布**] 门头沟区（东灵山，小龙门林场）。

[**保护策略**] 保护生境周围的原生植被，保持生态环境的稳定性。

Morphological Description: Rhizomes black-brown, tuberlike. Stems single, erect. Basal leaves forming a rosette, ovate-oblong to linear-oblong, base cuneate to rounded; stem leave only lowest leaves petiolate but frequently withering early, lanceolate to linear-lanceolate, apex obtuse to acuminate. Inflorescences branched. Calyx 5-parted nearly to base; lobes linear to linear-triangular. Corolla blue; limb slightly wider than tube, 5-lobed; lobes slightly spreading, nearly semiorbicular, margin entire. Nutlets wrinkled.

Distribution in Beijing: Mentougou District: Dongling Mountain, Xiaolongmen Forest Farm.

Protection Strategies: Protect the surrounding native vegetation of the habitat to maintain ecological stability.

056

中文名 **流苏树**

学　名 *Chionanthus retusus*

木樨科
Oleaceae

流苏树属
Chionanthus

[**形态描述**]　落叶灌木或乔木。叶片长圆形、椭圆形或圆形，先端圆钝，基部圆或宽楔形至楔形，中脉在上面凹入，下面凸起，侧脉 3～5 对。聚伞状圆锥花序；花萼 4 深裂，裂片尖三角形或披针形；花冠白色，裂片线状倒披针形。果椭圆形，被白粉，呈蓝黑色或黑色。花期 3～6 月，果期 6～11 月。

[**北京分布**]　房山区（上方山，霞云岭乡下石堡村）；门头沟区（妙峰山）；昌平区（十三陵）；延庆区（珍珠泉乡）；怀柔区（杏树台村）；密云区（新城子镇苏家峪村）。

[**保护策略**]　选择开阔地带，确保其生长环境有充足的光照；维护流苏树的原生生境，减少栖息地中的开发活动和环境破坏。

Morphological Description: Deciduous shrubs or tree; leaf blade oblong, elliptic, or orbicular, sometimes ovate or obovate, base rounded to cuneate, apex blunt; lateral veins 3–5 pairs. Cymose panicles; calyx 4-partite; lobes narrowly deltate or lanceolate. Corolla white; lobes linear-oblanceolate. Drupe blue-black or black, pruinose, ovoid. Fl. Mar–Jun, fr. Jun–Nov.

Distribution in Beijing: Fangshan District: Shangfang Mountain, Xiashibu Village of Xiayunling Township; Mentougou District: Miaofeng Mountain; Changping District: Ming Tombs; Yanqing District: Zhenzhuquan Township; Huairou District: Xingshutai Village; Miyun District: Sujiayu Village of Xinchengzi Township.

Protection Strategies: Ensure adequate sunlight in its growth environment. Maintain the native habitat of *Chionanthus retusus*, reducing development activities and environmental damage in habitats.

057

中文名 **珊瑚苣苔**

学　名 *Corallodiscus lanuginosus*

苦苣苔科
Gesneriaceae

珊瑚苣苔属
Corallodiscus

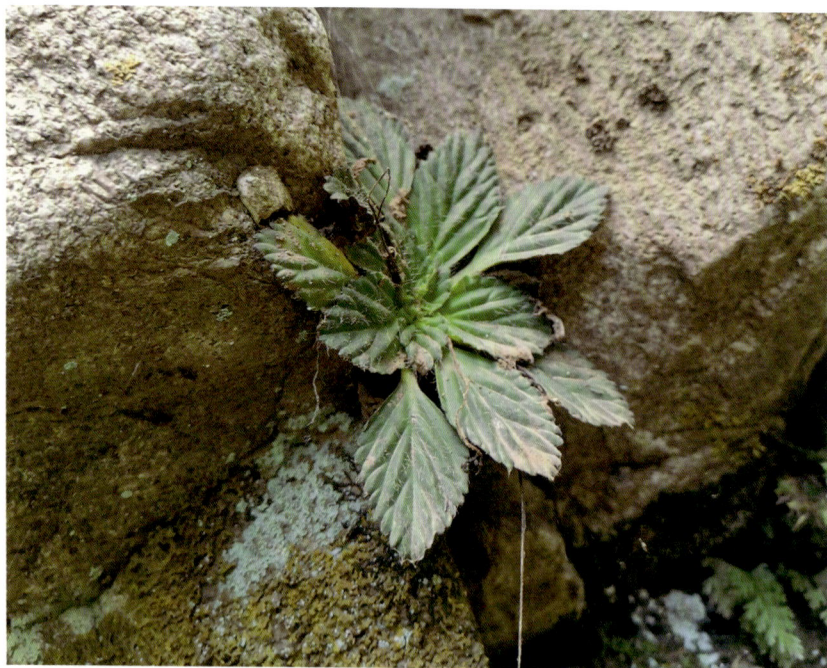

[**形态描述**] 多年生草本。叶全部基生，莲座状；叶片革质，倒披针形或卵状长圆形，顶端钝，基部楔形，边缘具细圆齿，上面明显泡状，疏被锈褐色长柔毛至近无毛，下面多为紫红色，近无毛，上面明显下凹，下面隆起；叶柄扁平，上面近无毛，下面密被锈色绵毛。聚伞花序 2～3 次分枝，每花序具 3～10 花；花萼 5 裂至近基部，裂片长圆形至长圆状披针形，外面疏被柔毛至无毛，内面无毛，具 3 脉；花冠筒状，二唇形，淡紫色、紫蓝色，外面无毛，内面下唇一侧具髯毛和斑纹；雄蕊 4，着生花冠基部；雌蕊无毛；子房长圆形，柱头头状，微凹。蒴果线形。花期 6 月，果期 8 月。

[**北京分布**] 房山区（百花山）。

[**保护策略**] 减少人类活动和干扰，保护其生长环境；维持栖息地内的生物多样性，增强生境的稳定性。

160

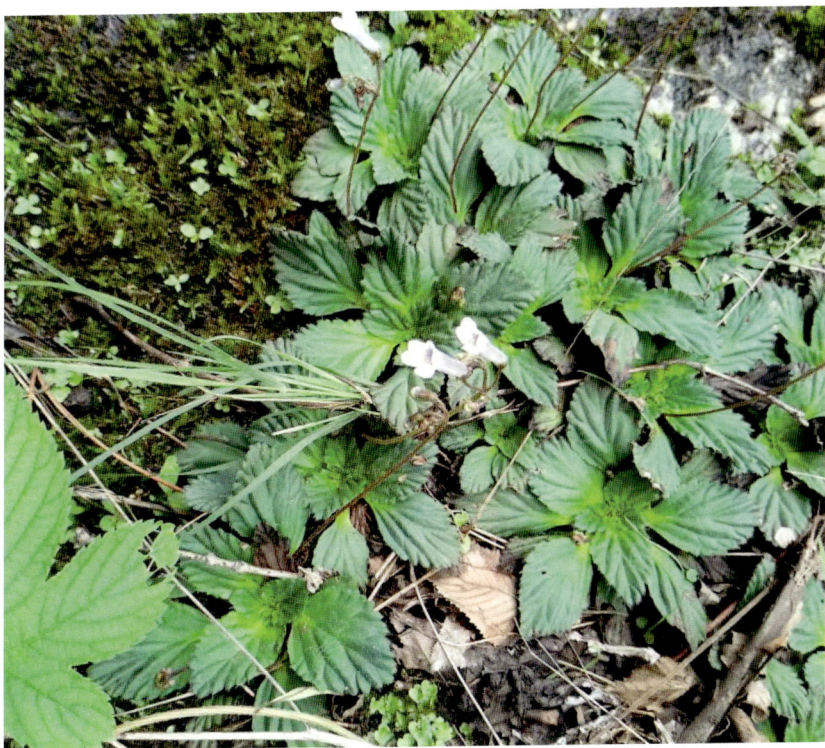

Morphological Description: Perennial herb. Leaves all basal, rosulate; leaf blade leathery, oblanceolate or ovate-oblong, obtuse at apex, cuneate at base, margin finely crenate, distinctly bullate above, sparsely rusty-brown villous to subglabrous, mostly purple-red beneath, subglabrous, concave above, convex beneath; petiole flattened, subglabrous above, densely rusty lanate beneath. Cymes 2–3-branched, each with 3–10 flowers; calyx 5-lobed nearly to base, lobes oblong to oblong-lanceolate, sparsely pubescent to glabrous outside, glabrous inside, 3-veined. Corolla tubular, bilabiate, pale purple to violet-blue, glabrous outside, inside with bearded maculae on one side of lower lip. Stamens 4, inserted at base of corolla. Pistil glabrous; ovary oblong; stigma capitate, slightly emarginate. Capsule linear. Fl. Jun, fr. Aug.

Distribution in Beijing: Fangshan District: Baihua Mountain.

Protection Strategies: Reduce human activities and disturbances, protect its growing environment. Maintain biodiversity in habitats, and enhance habitat stability.

058

中文名 **北玄参**

学　名 *Scrophularia buergeriana*

玄参科
Scrophulariaceae

玄参属
Scrophularia

[形态描述] 高大草本，高可达 1.5 米，根头肉质结节。茎四棱形，具白色髓心，略有自叶柄下延之狭翅。叶片卵形至椭圆状卵形，基部阔楔形至截形，边缘有锐锯齿。花序穗状，聚伞花序全部互生或下部的极接近而似对生，多少有腺毛；花萼裂片卵状椭圆形至宽卵形，顶端钝至圆形；花冠黄绿色，两唇的裂片均圆钝。蒴果卵圆形。花期 7 月，果期 8~9 月。

[北京分布] 门头沟区（东灵山）。

[保护策略] 减少人类活动和干扰，保护其生长环境；维持栖息地内的生物多样性，增强生境的稳定性。

Morphological Description: Herbs, to 1.5 m tall. Rhizomes with fleshy nodules. Stems narrowly winged, pith white. Leaf blade ovate to elliptic-ovate, base broadly cuneate to truncate, margin acutely serrate. Thyrses spicate; cymes all alternate or lower ones subopposite. Sparsely glandular hairy. Lobes ovate-elliptic to broadly ovate. Corolla yellow-green; lip lobes rounded to obtuse. Capsule ovoid. Fl. Jul, fr. Aug–Sep.

Distribution in Beijing: Mentougou District: Dongling Mountain.

Protection Strategies: Reduce human activities and disturbances, protect its growing environment. Maintain biodiversity in habitats, and enhance habitat stability.

059 中文名 **玄参**

学　名 *Scrophularia ningpoensis*

玄参科
Scrophulariaceae

玄参属
Scrophularia

[形态描述] 高大草本，可达1.5米余。支根数条，纺锤形或胡萝卜状膨大。茎四棱形，无翅或有极狭的翅，无毛或多少有白色卷毛，常分枝。叶对生，上部有时互生，叶片多为卵形，有时上部的为卵状披针形至披针形，基部楔形、圆形或近心形，边缘具细锯齿。花序为疏散的大圆锥花序；花萼裂片圆形，边缘稍膜质；花冠筒多少球形，裂片圆形。蒴果卵圆形。花期6～10月，果期9～11月。

[北京分布] 密云区；海淀区（金山）。

[保护策略] 在原生地中，避免过度采摘和人为干扰。

Morphological Description: Herbs, to 1.5 m tall. Lateral roots few, fusiform to conical. Stems quadrangular, lightly grooved to nearly winged, glabrous or white crisped hairy. Leaves opposite, sometimes apically alternate; leaf blade below mostly ovate, upper ones ovate-lanceolate to lanceolate, base cuneate, rounded, or subcordate, margin serrulate. Thyrses largely lax; cymes terminal and axillary; lobes suborbicular, rounded, margin submembranous. Corolla tube subglobose; lobes orbicular. Capsule ovoid. Fl. Jun–Oct, fr. Sep–Nov.

Distribution in Beijing: Miyun District; Haidian District: Jinshan Mountain.

Protection Strategies: In the native habitat, avoid excessive harvesting and human interference.

060

中 文 名 **羊乳**

学　　名 *Codonopsis lanceolata*

桔梗科
Campanulaceae

党参属
Codonopsis

[形态描述] 植株全体光滑无毛。茎基略近于圆锥状，根常肥大呈纺锤状。茎缠绕，黄绿而微带紫色。叶在主茎上互生，披针形或菱状狭卵形；在小枝顶端通常2～4叶簇生，叶片菱状卵形、狭卵形或椭圆形，通常全缘或有疏波状锯齿，上面绿色，下面灰绿色。花单生或对生于小枝顶端；花萼贴生，裂片卵状三角形；花冠阔钟状，裂片黄绿色或乳白色内有紫色斑。蒴果下部半球状；种子多数，卵形，有翼，棕色。花果期7～8月。

[北京分布] 门头沟区（小龙门林场，东灵山）；怀柔区（南苇滩，喇叭沟门，琉璃庙）；房山区（地藏庵，上方山）；昌平区（十三陵，辛庄村）；密云区（半城子村，豹子岭，遥桥峪）；平谷区（东指壶）；延庆区（兰角沟）。

[保护策略] 减少人类采挖，保护其生长环境；维持栖息地内的生物多样性，增强生境的稳定性。

Morphological Description: Plants glabrous throughout or occasionally sparsely villous on stems and leaves. Caudexes subcylindrical. Roots usually fusiform-thickened. Stems twining, yellow-green but with purplish shade. Leaves on main stems alternate, lanceolate, ovate, or elliptic; usually leaves 2–4-fascicled on top of branchlets, subopposite or verticillate; blade abaxially gray-green, adaxially green, ovate, narrowly ovate, or elliptic, margin usually entire or sparsely sinuate. Flowers solitary or paired on top of branchlets. Calyx adnate to ovary by half; lobes ovate or deltoid. Corolla broadly campanulate; lobes yellow-green or milk-white, with purple spots. Capsule hemispherical at base. Seeds numerous, brown, winged. Fl. and fr. Jul–Aug.

Distribution in Beijing: Mentougou District: Xiaolongmen Forest Farm, Dongling Mountain; Huairou District: Nanweitan, Labagoumen, Liulimiao; Fangshan District: Dizang Temple, Shangfang Mountain; Changping District: Ming Tombs, Xinzhuang Village; Miyun District: Banchengzi Village, Baoziling, Yaoqiaoyu; Pinggu District: Dongzhihu Mountain; Yanqing District: Lanjiaogou.

Protection Strategies: Reduce human excavation, protect its growing environment. Maintain biodiversity in habitats, and enhance habitat stability.

061

中文名 **睡菜**

学　名 *Menyanthes trifoliata*

睡菜科
Menyanthaceae

睡菜属
Menyanthes

[形态描述] 多年生沼生草本。匍匐状根状茎粗大。叶全部基生，挺出水面，三出复叶，小叶椭圆形，先端钝圆，基部楔形，全缘或边缘微波状，中脉明显；总状花序多花；苞片卵形，先端钝，全缘；裂片卵形，先端钝；花冠白色，筒形，内面具白色长流苏状毛，裂片椭圆状披针形，花丝扁平，线形。蒴果球形；种子臌胀，圆形，表面平滑，花果期5~7月。

[北京分布] 延庆区（蔡家河，张山营）。

[保护策略] 加强生境保护，避免水位大幅波动，并维持水质；维持生境内植物群落平衡，避免单一物种过度繁衍而抑制睡菜生存，进而导致其消失或灭绝。

Morphological Description: Perennial marsh herb. Creeping rhizomes thick. Leaves all basal, emergent, trifoliolate; leaflets elliptic, apex obtuse-round, base cuneate, margin entire or undulate, midvein distinct. Racemes many-flowered; bracts ovate, apex obtuse, margin entire. Lobes ovate, apex obtuse. Corolla white, tubular, inner surface with long fimbriate white hairs; lobes elliptic-lanceolate. Filaments flat, linear. Capsules globose. Seeds swollen, orbicular, surface smooth. Flowering and fruiting period May–July.

Distribution in Beijing: Yanqing District: Caijiahe, Zhangshanying.

Protection Strategies: Strengthen habitat protection, avoid significant fluctuations in water levels, and maintain water quality. Maintain the balance of plant communities within the habitat, prevent any single species from over-proliferating and suppressing the survival of *Menyanthes trifoliata*, thereby avoiding its disappearance or extinction.

062

中文名 **大头风毛菊**
学　名 *Saussurea baicalensis*

菊科
Asteraceae

风毛菊属
Saussurea

[形态描述] 多年生草本，高30～60厘米。茎直立。叶椭圆状披针形，顶端急尖或渐尖，基部楔形，边缘尖锯齿，两面绿色，粗糙，被长柔毛。头状花序多数，沿茎成紧密的总状花序；总苞钟状；总苞片3～4层，披针形，顶端渐尖或急尖；小花紫色。瘦果长圆形；冠毛白色。花果期6～7月。

[北京分布] 门头沟区（东灵山，江水河村，小龙门）。

[保护策略] 避免高山草甸和裸岩地区的开发活动，保持土壤的原生特性。

Morphological Description: Perennial herb, 30–60 cm tall. Stem erect. Leaves elliptic-lanceolate, apex acute or acuminate, base cuneate, margin with acute serrate teeth, both surfaces green, scabrous, and villous. Numerous capitula arranged in dense racemes along stem; involucre campanulate; phyllaries 3–4-seriate, lanceolate, apex acuminate or acute; florets purple. Achenes oblong; pappus white. Fl. and fr. Jun–Jul.

Distribution in Beijing: Mentougou District: Dongling Mountain, Jiangshuihe Village, Xiaolongmen.

Protection Strategies: Avoid development activities in alpine meadows and bare rock areas, and preserve the original characteristics of the soil.

063

中 文 名 **款冬**

学　名 *Tussilago farfara*

菊科
Asteraceae

款冬属
Tussilago

[形态描述] 多年生草本。根状茎横生地下，褐色。早春花叶抽出数个花葶，密被白色茸毛，有鳞片状、互生的苞叶，苞叶淡紫色；后生出的基生叶阔心形，叶片边缘有波状、顶端增厚的疏齿，掌状网脉，下面被密白色茸毛。头状花序单生顶端；总苞片1~2层，总苞钟状，总苞片线形，常带紫色，被白色柔毛及腺毛；边缘有多层雌花，黄色；中央的两性花少数，花冠管状；柱头头状。瘦果圆柱形；冠毛白色。

[北京分布] 延庆区（北京世园公园）；昌平区（黑山寨，辛庄，流村镇）；门头沟区（东灵山）。

[保护策略] 在其栖息地内，尽量避免过度开垦、采矿等活动；合理调控栖息地植被多样性，以确保款冬和其他植物的共生关系；适当轮作或混种，防止单一植株聚集引发病虫害。

Morphological Description: Perennial herb. Rhizomes long creeping, subterranean, brown. Densely white lanate, with scale-shaped alternate purple-violet bracteate leaves. Basal leaves orbicular-cordate, abaxially densely white tomentose, palmately reticulate veined, margin undulate, unequally toothed. Capitula solitary, terminal. Involucres campanulate; phyllaries 1–2-seriate, linear, white villous, often purple tinged. Ray florets female, many seriate, yellow. Disk florets few; stigma capitate. Achenes cylindric. Pappus white.

Distribution in Beijing: Yanquing District: Beijing Expo Park; Changping District: Heishanzhai, Xinzhuang, Liucun Town; Mentougou District: Dongling Mountain.

Protection Strategies: In its habitat, avoid excessive reclamation, mining, and similar activities as much as possible. Control vegetation diversity in the habitat to ensure the symbiotic relationship between *Tussilago farfara* and other plants. Practice appropriate crop rotation or intercropping to prevent pest and disease outbreaks caused by monoculture.

064

中文名 **楤木**
学　名 *Aralia elata*

五加科
Araliaceae

楤木属
Aralia

[**形态描述**] 灌木或乔木，高2~5米；小枝疏生细刺。叶为二回或三回羽状复叶；羽片有小叶5~11，基部有小叶1对；小叶片卵形、阔卵形或长卵形，先端渐尖或短渐尖，基部圆形，边缘有锯齿。圆锥花序大，分枝密生淡黄棕色或灰色短柔毛；苞片锥形，外面有毛；花白色，芳香；萼无毛，边缘有5个三角形小齿；花瓣5，卵状三角形。果实球形；宿存花柱离生或合生至中部。花期7~9月，果期9~12月。

[**北京分布**] 怀柔区（百泉山，慕田峪）；房山区（上方山，十渡）；密云区（五座楼）。

[**保护策略**] 优先保护林缘、林中空地和山坡等具有湿润、肥沃土壤条件的微生境，严格管控森林砍伐和土地开发等行为，避免对楤木栖息地造成破坏；通过植被覆盖等措施保持土壤养分；维持栖息地适度的透光率，保障楤木生长所需的光照条件。

Morphological Description: Shrubs or small trees, 2–5 m tall. Branches armed with sparse prickles. Leaves 2(or 3)-pinnately compound; leaflets 5–11 per pinna, broadly ovate to elliptic-ovate or narrowly ovate, base cordate to subcordate or rounded, margin serrulate, apex acuminate. Inflorescence a terminal panicle of umbels, densely yellow-brown or gray pubescent; bracts persistent, lanceolate to subulate, sometimes ciliate. Fruit globose; styles persistent. Fl. Jul–Sep, fr. Sep–Dec.

Distribution in Beijing: Huairou District: Baiquan Mountain, Mutianyu; Fangshan District: Shangfang Mountain, Shidu; Miyun District: Wuzuolou.

Protection Strategies: Prioritize the protection of moist, fertile soil environments such as forest edges, clearings, and mountain slopes; prevent habitat destruction caused by deforestation and land development. Maintain soil nutrients through measures like vegetation coverage. Sustain moderate light transmittance in habitats to ensure the light conditions required for *Aralia elata* growth.

065

中文名 **刺五加**
学　名 *Eleutherococcus senticosus*

五加科
Araliaceae

五加属
Eleutherococcus

[形态描述] 灌木，高 1～6 米；分枝多，通常密生刺；叶有小叶 5；叶柄常疏生细刺；小叶片椭圆状倒卵形或长圆形，先端渐尖，基部阔楔形，边缘有锐利重锯齿，侧脉 6～7 对，两面明显。伞形花序单个顶生，或 2～6 个组成稀疏的圆锥花序，有花多数；花梗无毛或基部略有毛；花紫黄色；萼无毛，边缘近全缘；子房 5 室。果实球形或卵球形。花期 6～7 月，果期 8～10 月。

[北京分布] 房山区（百花山，霞云岭）；平谷区（北水峪）；门头沟区（小龙门，东灵山）；延庆区（松山，海坨山）；密云区（坡头）；怀柔区（慕田峪）。

[保护策略] 优先保护原生林地及山坡湿润生境；严格管控过度砍伐，避免因林冠层破坏导致生境光照过强；通过维持群落物种多样性，防止生境内单优势种群过度扩张而挤占刺五加的光照和水分资源。

Morphological Description: Shrubs, 1–6 m tall. Branches with dense to scattered, slender, terete, bristlelike prickles; leaflets 5, elliptic-obovate or oblong, secondary veins 6–7 pairs, conspicuous on both surfaces, base broadly cuneate, margin sharply biserrate, apex shortly acuminate or acuminate. Inflorescence terminal, a solitary or compound umbel, usually with 2–6 umbels together; pedicels glabrous or slightly pubescent at base. Calyx subentire. Corolla purple-yellow Ovary 5-carpellate. Fruit ovoid-globose. Fl. Jun–Jul, fr. Aug–Oct.

Distribution in Beijing: Fangshan District: Baihua Mountain, Xiayunling; Pinggu District: Beishuiyu; Mentougou District: Xiaolongmen, Dongling Mountain; Yanqing District: Songshan Mountain, Haituo Mountain; Miyun District: Potou; Huairou District: Mutianyu.

Protection Strategies: Priority should be given to protecting native forest lands and moist hillslope habitats. Strictly regulate excessive logging to avoid excessive light intensity in habitats caused by canopy layer damage. Maintain community species diversity to prevent overexpansion of single dominant populations within habitats from competing for light and water resources of *Eleutherococcus senticosus*.

066

中文名 **无梗五加**

学　名 *Eleutherococcus sessiliflorus*

五加科
Araliaceae

五加属
Eleutherococcus

[**形态描述**] 灌木或小乔木，高2～5米；枝灰色，无刺或疏生刺。叶柄无刺或有小刺；小叶片倒卵形或长圆状倒卵形至长圆状披针形，先端渐尖，基部楔形，两面均无毛，边缘有不整齐锯齿。头状花序紧密；总花梗密生短柔毛；萼密生白色绒毛，边缘有5小齿；花瓣5，卵形，浓紫色；子房2室，花柱全部合生成柱状，柱头离生。果实倒卵状椭圆球形。花期8～9月，果期9～10月。

[**北京分布**] 门头沟区（九龙山，小龙门林场，妙峰山）；延庆区（大庄科乡，西沟里，兰角沟，玉渡山，松山）；密云区（云蒙山，遥桥峪，花园村）；怀柔区（喇叭沟门，龙潭沟，苗营，黑坨山，琉璃庙）；房山区（霞云岭）。

[**保护策略**] 防止伐木和开发行为破坏其自然生长环境；保持疏林生境，避免过度砍伐造成阳光直射；保护其栖息地时应维持生境内的植物多样性，以增强生态系统的稳定性。

Morphological Description: Shrubs or small trees, 2–5 m tall. Branches gray, unarmed or sparsely spiny. Petiole unarmed or with small spines; leaflets obovate or oblong-obovate to oblong-lanceolate, apex acuminate, base cuneate, both surfaces glabrous, margin irregularly serrate. Capitula dense; peduncles densely pubescent. Calyx densely white tomentose, margin with 5 small teeth; petals 5, ovate, dark purple. Ovary 2-loculed; styles completely united into a column, stigmas free. Fruit obovoid-ellipsoid. Fl. Aug–Sep, fr. Sep–Nov.

Distribution in Beijing: Mentougou District: Jiulong Mountain, Xiaolongmen Forest Farm, Miaofeng Mountain; Yanqing District: Dazhuangke Township, Xigouli, Lanjiaogou, Yudu Mountain, Songshan Mountain; Miyun District: Yunmeng Mountain, Yaoqiaoyu, Huayuan Village; Huairou District: Labagoumen, Longtangou, Miaoying, Heituo Mountain, Liulimiao; Fangshan District: Xiayunling.

Protection Strategies: Prevent logging and development activities that may damage its natural growing environment. Maintain open wooded habitats to avoid excessive deforestation that could lead to direct sunlight exposure. When protecting its habitat, maintain plant diversity within the habitat to enhance the stability of the ecosystem.

067

中文名 **刺楸**

学　名 *Kalopanax septemlobus*

五加科
Araliaceae

刺楸属
Kalopanax

[形态描述] 落叶乔木；小枝淡黄棕色或灰棕色，散生粗刺。叶片在长枝上互生，在短枝上簇生，圆形或近圆形，掌状5～7浅裂，裂片阔三角状卵形至长圆状卵形，先端渐尖，基部心形，边缘有细锯齿；叶柄细长，无毛。圆锥花序大；总花梗无毛；花白色或淡绿黄色；萼无毛，边缘有5小齿；花瓣5，三角状卵形；雄蕊5；子房2室。果实球形，蓝黑色。花期7～10月，果期9～12月。

[北京分布] 延庆区。

[保护策略] 避免土地开发和农业扩展导致栖息地丧失；防止水源污染，保持土壤的湿润性；避免过度遮阴，以确保其获得充足的光照；维持栖息地内的植物多样性，以增强生态系统的稳定性和抵御力。

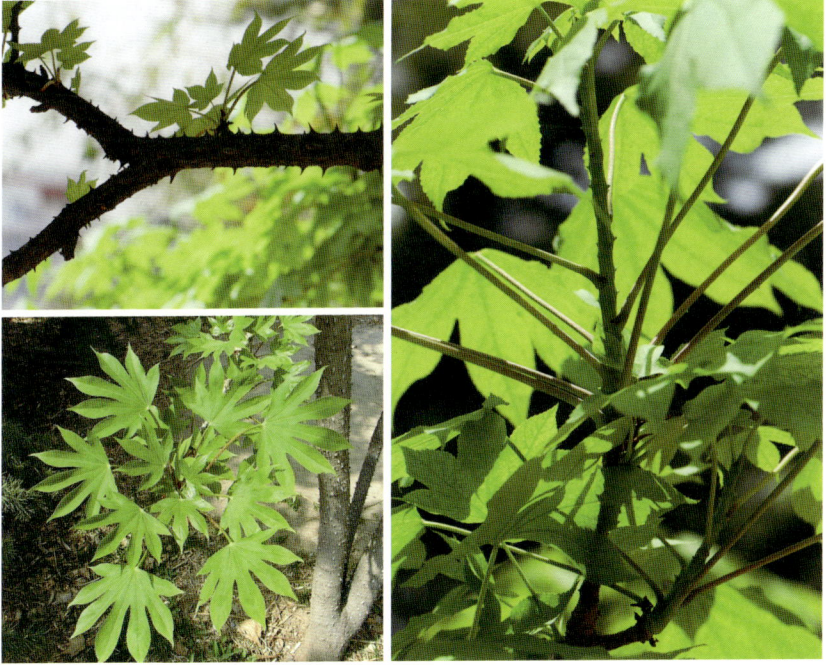

Morphological Description: Deciduous trees; branchlets yellowish-brown or gray-brown, sparsely armed with thick spines. Leaves alternate on long shoots, clustered on short shoots, circular or suborbicular, palmately 5–7 shallowly lobed; lobes broadly triangular-ovate to oblong-ovate, apex acuminate, base cordate, margin finely serrate; petioles slender, glabrous. Large panicles; peduncles glabrous; flowers white or pale greenish-yellow; calyx glabrous, margin with 5 small teeth; petals 5, triangular-ovate; stamens 5; ovary 2-loculed. Fruit spherical, blue-black. Fl. Jul–Oct, fr. Sep–Dec.

Distribution in Beijing: Yanqing District.

Protection Strategies: Avoid land development and agricultural expansion that lead to habitat loss. Prevent water source pollution and maintain soil moisture; avoid excessive shading to ensure sufficient sunlight. Maintain plant diversity in its habitat to enhance ecosystem stability and resilience.

068

中文名 **凹舌兰**

学　名 *Dactylorhiza viridis*

兰科
Orchidaceae

掌裂兰属
Dactylorhiza

[**形态描述**]　植株高14～45厘米。块茎肉质，前部呈掌状分裂。茎直立，基部具筒状鞘。叶片椭圆形或椭圆状披针形，先端钝或急尖。总状花序，花苞片线形或狭披针形；花绿黄色或绿棕色；中萼片直立，凹陷呈舟状；侧萼片偏斜，卵状椭圆形；花瓣线状披针形；唇瓣肉质，倒披针形，基部具囊状距。蒴果直立，椭圆形，无毛。花期6～8月，果期9～10月。

[**北京分布**]　门头沟区（百花山）；延庆区（海坨山）；密云区（雾灵山）。

[**保护策略**]　避免城市化、农业扩展或开发活动导致栖息地丧失；防止土壤和水源的污染；维持生物多样性，提升土壤质量和栖息地稳定性；避免在凹舌兰的生长区域引入外来物种，以防止生态失衡。

Morphological Description: Plants 14–45 cm tall. Tuber fleshy, palmately divided at front. Stem erect, with tubular sheaths at base. Leaves elliptic or elliptic-lanceolate, apex obtuse or acute. Raceme; bracts linear or narrowly lanceolate; flowers greenish yellow or greenish brown. Dorsal sepal erect, concave and navicular; lateral sepals oblique, ovate-elliptic. Petals linear-lanceolate. Lip fleshy, oblanceolate, with saccate spur at base. Capsule erect, elliptic, glabrous. Fl. Jun–Aug, fr. Sep–Oct.

Distribution in Beijing: Mentougou District: Baihua Mountain; Yanqing District: Haituo Mountain; Miyun District: Wuling Mountain.

Protection Strategies: Avoid habitat loss caused by urbanization, agricultural expansion, or development activities. Prevent soil and water pollution. Maintain plant diversity, improve soil quality, and enhance habitat stability. Avoid introducing invasive species into the growing areas of *Dactylorhiza viridis* to prevent ecological imbalance.

069

中 文 名 **蜻蜓兰**
学　　名 *Platanthera souliei*

兰科
Orchidaceae

舌唇兰属
Platanthera

[形态描述]　植株高20～60厘米。根状茎指状。茎下部的2（～3）枚大叶片倒卵形或椭圆形，在大叶之上具1至几枚苞片状小叶。总状花序；花苞片狭披针形；花黄绿色；中萼片凹陷呈舟状，卵形；侧萼片斜椭圆形；花瓣斜椭圆状披针形，与中萼片相靠合，稍肉质；唇瓣舌状披针形，肉质，侧裂片三角状镰形，先端锐尖；距细长，下垂。花期6～8月，果期9～10月。

[北京分布]　房山区（蒲洼）；门头沟区（百花山）；怀柔区（喇叭沟门）；密云区（坡头）。

[保护策略]　维持栖息地内植物多样性，促进生态系统的稳定性；通过人为手段调控林分结构，保持适当的荫蔽和湿度，促进目标物种种群扩繁。

Morphological Description: Plants 20–60 cm tall. Rootstock digitate. Lower stem with 2(–3) large obovate or elliptic leaves, above which are 1 to several bract-like small leaves. Raceme; floral bracts narrowly lanceolate. Flowers yellowish-green. Dorsal sepal concave-navicular, ovate; lateral sepals obliquely elliptic. Petals obliquely elliptic-lanceolate, connivent with dorsal sepal, slightly fleshy; lip ligulate-lanceolate, fleshy, lateral lobes triangular-falcate, apex acute; spur slender, pendulous. Fl. Jun–Aug, Fr. Sep–Oct.

Distribution in Beijing: Fangshan District: Puwa; Mentougou District: Baihua Mountain; Huairou District: Labagoumen; Miyun District: Potou.

Protection Strategies: Maintain plant diversity in the habitat to promote ecosystem stability. Regulate stand structure through anthropogenic means, preserve appropriate shade and humidity levels, and facilitate population expansion and reproduction of the target species.

070

中 文 名 **小花蜻蜓兰**
学　　名 *Platanthera ussuriensis*

兰科
Orchidaceae

舌唇兰属
Platanthera

[形态描述] 植株高20～55厘米。根状茎指状，肉质。茎基部具1～2枚筒状鞘，下部的2～3枚大叶片匙形或狭长圆形。总状花序；花苞片狭披针形；子房细圆柱形；中萼片凹陷呈舟状，宽卵形；侧萼片张开或反折，狭椭圆形；花瓣直立，狭长圆状披针形，与中萼片相靠合，肉质；唇瓣向前伸展，舌状披针形，基部两侧各具1枚小侧裂片，中裂片舌状；距纤细，下垂。花期7～8月，果期9～10月。

[北京分布] 密云区（小西天，坡头）；门头沟区（清水林场，妙峰山）。

[保护策略] 保留周围的植被以减少水分蒸发，创造适合其生长的微气候条件；小花蜻蜓兰通常附生在树木上，应确保其附生环境的稳定。

Morphological Description: Plants 20–55 cm tall. Rootstock digitate and fleshy. Stem with 1–2 tubular sheaths at base, lower 2 3 leaves spatulate or narrowly oblong. Raceme; floral bracts narrowly lanceolate; ovary slenderly cylindric. Dorsal sepal concave-navicular, broadly ovate; lateral sepals spreading or reflexed, narrowly elliptic. Petals erect, narrowly oblong-lanceolate, connivent with dorsal sepal, fleshy; lip forward-extending, ligulate-lanceolate, with 1 small lateral lobe on each side at base, mid-lobe ligulate; spur slender, pendulous. Fl. Jul–Aug, Fr. Sep–Oct.

Distribution in Beijing: Miyun District: Xiaoxitian, Potou; Mentougou District: Qingshui Forest Farm, Miaofeng Mountain.

Protection Strategies: Retain surrounding vegetation to reduce water evaporation and create microclimatic conditions suitable for its growth. *Platanthera ussuriensis* typically grows as an epiphyte on trees, so ensure the stability of its epiphytic environment.

071

中文名 **二叶舌唇兰**
学　名 *Platanthera chlorantha*

兰科
Orchidaceae

舌唇兰属
Platanthera

[形态描述]　植株高 30～50 厘米。块茎卵状纺锤形，肉质。茎直立，无毛，近基部具 2 枚彼此紧靠、近对生的大叶。基部大叶片椭圆形，基部收狭成抱茎的鞘状柄。总状花序具 12～32 朵花；花苞片披针形；花较大，绿白色或白色；中萼片舟状，圆状心形；侧萼片张开，斜卵形；花瓣偏斜，狭披针形；唇瓣舌状，肉质；距棒状圆筒形；蕊柱粗，药室明显叉开；蕊喙宽。花期 6～7（～8）月。

[北京分布]　门头沟区；密云区；延庆区。

[保护策略]　避免森林砍伐和不合理开发，以维持其生长所需的微环境；保持周围植被的完整性，以减少水分蒸发；选择生长健康、稳定的树木作为其附生基质；对野生植株的采集实施严格限制；维持栖息地中植物多样性，提高生态系统的稳定性。

Morphological Description: Plants 30–50 cm tall. Rootstock tuberous, ovoid-fusiform. Stem erect, with a tubular sheath at base, 2-leaved. Leaves basal, subopposite, spatulate-elliptic or oblanceolate-elliptic, gradually contracted and sheathing at base; Raceme with 12–32 flowers; bracts lanceolate. Flowers large, greenish white or white. Dorsal sepal Cymbiform, orbicular-cordate, cymbiform; lateral sepals spreading, ovate, oblique. Petals narrowly ovate-lanceolate, oblique; lip spreading to pendulous, ligulate; spur clavate-cylindric. Column stout; anther locules divergent; rostellum broad. Fl. Jun–Jul(–Aug).

Distribution in Beijing: Mentougou District; Miyun District; Yanqing District.

Protection Strategies: Avoid deforestation and unreasonable development to maintain the microenvironment in which the plants grow. Preserve the integrity of surrounding vegetation to reduce moisture evaporation; select healthy, stable trees as the substrate for epiphytic growth. Strictly restrict the collection of wild plants. Maintain plant diversity in the habitat to improve the stability of the ecological environment.

072 中文名 **尖唇鸟巢兰**
学 名 *Neottia acuminata*

兰科
Orchidaceae

鸟巢兰属
Neottia

[形态描述] 植株高14～30厘米。茎直立，无毛，中部以下具3～5枚鞘；鞘膜质。总状花序顶生，通常具20余朵花；花序轴无毛；花苞片长圆状卵形，无毛；子房椭圆形，无毛；花小，黄褐色，常3～4朵聚生而呈轮生状；中萼片狭披针形，无毛；侧萼片与中萼片相似；花瓣狭披针形；唇瓣形状变化较大；蕊柱极短；柱头横长圆形，直立，左右两侧内弯，围抱蕊喙。蒴果椭圆形。花果期6～8月。

[北京分布] 门头沟区（百花山，东灵山）；延庆区（海坨山）；密云区（雾灵山）。

[保护策略] 避免过度开发、森林砍伐等活动，确保其附生环境的稳定；防止过度采集造成种群减少。

Morphological Description: Plants 14–30 cm tall. Stem erect, glabrous, with 3–5 sheaths below middle; sheaths membranous. Terminal raceme, usually with more than 20 flowers; rachis glabrous; floral bracts oblong-ovate, glabrous; ovary elliptic, glabrous. Flowers small, yellowish-brown, often 3–4 clustered and verticillate. Dorsal sepal narrowly lanceolate, glabrous; lateral sepals similar to dorsal sepal. Petals narrowly lanceolate; lip highly variable in shape. Column extremely short; stigma transversely oblong, erect, incurved on both sides and embracing rostellum. Capsule elliptic. Fl. and fr. Jun–Aug.

Distribution in Beijing: Mentougou District: Baihua Mountain, Dongling Mountain; Yanqing District: Haituo Mountain; Miyun District: Wuling Mountain.

Protection Strategies: Avoid excessive development and deforestation to ensure the stability of its epiphytic environment. Prevent over-collection that causes population decline.

073

中文名 **十字兰**
学　名 *Habenaria schindleri*

兰科
Orchidaceae

玉凤花属
Habenaria

[**形态描述**] 植株高25～70厘米。块茎肉质。茎直立，具多枚疏生的叶。中下部的叶片线形，基部成抱茎的鞘。总状花序具10～20余朵花；花苞片线状披针形，无毛；子房圆柱形，扭转，无毛；花白色，无毛；中萼片卵圆形，直立，凹陷呈舟状，与花瓣靠合呈兜状；侧萼片强烈反折，斜长圆状卵形；花瓣轮廓近三角形，2裂，上裂片先端稍钝，下裂片小齿状，三角形，先端二浅裂；唇瓣向前伸，基部线形，近基部的1/3处3深裂，呈十字形，中裂片全缘，侧裂片与中裂片垂直伸展，向先端增宽且具流苏；距下垂，近末端突然膨大，向前弯曲，末端钝；柱头2个。花期7～9（～10）月。

[**北京分布**] 门头沟区（东灵山）。

[**保护策略**] 限制野生植株的采集；维持其栖息地内植物多样性，增强生态系统稳定性。

Morphological Description: Plants 25–70 cm tall. Tubers fleshy. Stem erect, with several scattered leaves. The leaves in the lower and middle parts of the stem are linear, with a sheathing base that clasps the stem. The racemose inflorescence contains 10–20 flowers; floral bracts lanceolate and glabrous. Ovary cylindrical, twisted, glabrous. Flowers white, glabrous. Dorsal sepal ovate, erect, concave, boat-shaped, and adaxially fused with the petals to form a hood. Lateral sepals strongly reflexed, obliquely ovate-lanceolate. Petals triangular, semi-equilateral, deeply 2-lobed; upper lobe with a slightly blunt apex; lower lobe small, serrate, triangular, with two shallow lobes at the apex. Lip extending forward, linear at the base, 3-deeply lobed at the proximal 1/3 of the base, forming a cruciform shape; median lobe entire; lateral lobes perpendicular to the median lobe, widening toward the apex, with tasseled tips. Spur pendent, abruptly enlarged near the apex, bending forward, and blunt at the tip. Column with two stigmas. Fl. Jul–Sep(–Oct).

Distribution in Beijing: Mentougou District: Dongling Mountain.

Protection Strategies: Restrict the collection of wild plants. Maintain plant diversity within the habitat to enhance ecosystem stability.

074

中文名 **珊瑚兰**

学　名 *Corallorhiza trifida*

兰科
Orchidaceae

珊瑚兰属
Corallorhiza

[形态描述] 腐生小草本，高 10 ~ 22 厘米；根状茎肉质，珊瑚状。茎直立，红褐色，无绿叶，被 3 ~ 4 枚鞘；鞘圆筒状，红褐色。总状花序；花苞近长圆形；花淡黄色或白色；中萼片狭长圆形，先端钝或急尖；侧萼片与中萼片相似；花瓣近长圆形，与中萼片靠合成盔状；唇瓣近长圆形，3 裂；侧裂片较小；中裂片近椭圆形；唇盘上有 2 条肥厚的纵褶片；蕊柱较短，两侧具翅。蒴果下垂，椭圆形。花果期 6 ~ 8 月。

[北京分布] 门头沟区（百花山，东灵山）；延庆区（海坨山）。

[保护策略] 保持寄主树木健康；维持栖息地内植物多样性，增强生态系统稳定性；使生境保持适度湿润，但避免积水。

Morphological Description: Saprophytic herbs, 10–22 cm tall; rhizome fleshy, coral-like. Stem erect, reddish-brown, without green leaves, covered by 3–4 sheaths; sheaths cylindrical, reddish-brown. Inflorescence a raceme; floral buds oblong-elliptic; flowers pale yellow or white; dorsal sepal narrowly oblong, with an obtuse or acute apex; lateral sepals similar to the dorsal sepal; petals narrowly oblong, forming a hood-like structure with the dorsal sepal; lip petals narrowly oblong, 3-lobed; lateral lobes smaller; median lobe nearly elliptical; lip disc with two thickened longitudinal ridges; column short, with wings on both sides. Capsule pendent, ellipsoid. Fl. and fr. period: Jun–Aug.

Distribution in Beijing: Mentougou District: Baihua Mountain, Dongling Mountain; Yanqing District: Haituo Mountain.

Protection Strategies: Maintain the health of host trees. Preserve plant diversity within the habitat to enhance ecosystem stability. Keep the habitat moderately moist while avoiding waterlogging.

075

中文名 **二叶兜被兰**
学　名 *Neottianthe cucullata*

兰科
Orchidaceae

兜被兰属
Neottianthe

[形态描述] 植株高4~24厘米。块茎圆球形。茎直立，基部具1~2枚圆筒状鞘，其上具2枚近对生的叶，在叶之上常具1~4枚披针形的不育苞片。叶片卵形，先端急尖或渐尖，基部骤狭成抱茎的短鞘，叶上面有时具紫红色斑点。总状花序，花苞片披针形；子房圆柱状纺锤形；花紫红色或粉红色；萼片彼此紧密靠合成兜状；侧萼片斜镰状披针形，先端急尖；花瓣披针状线形，与萼片贴生；唇瓣上面和边缘具细乳突，中部3裂，侧裂片线形，先端急尖，具1脉，中裂片先端钝，具3脉；距细圆筒状圆锥形，中部向前弯曲。花期8~9月。

[北京分布] 房山区（上方山）；门头沟区（百花山，东灵山，小龙门）；延庆区（玉渡山）；怀柔区（箭扣）；密云区（锥峰山）。

[保护策略] 限制野生植株的采集；保持栖息地湿度和温度相对稳定；维持栖息地内植物多样性，增强生态系统稳定性。

Morphological Description: Plants 4–24 cm tall. Tuber spherical. Stem erect, with 1–2 cylindrical sheaths at the base, bearing 2 nearly opposite leaves, and often 1–4 lanceolate infertile bracts above the leaves. Leaves ovate, with an acute or gradually tapering apex, base abruptly narrowed into a short sheath clasping the stem, sometimes with purple-red spots on the upper surface. Inflorescence a raceme, with lanceolate floral bracts; ovary cylindric-fusiform; flowers purple-red or pink; sepals closely adpressed to form a hood; lateral sepals obliquely falcate-lanceolate, with an acute apex; petals lanceolate-linear, adpressed to the sepals; lip with fine papillae on the surface and margins, 3-lobed in the center; lateral lobes linear, with an acute apex and 1 vein; median lobe with a blunt apex and 3 veins; spur slender, cylindric-conical, bending forward at the middle. Fl. Aug–Sep.

Distribution in Beijing: Fangshan District: Shangfang Mountain; Mentougou District: Baihua Mountain, Dongling Mountain, Xiaolongmen; Yanqing District: Yudu Mountain; Huairou District: Jiankou; Miyun District: Zhuifeng Mountain.

Protection Strategies: Restrict the collection of wild plants. Maintain relatively stable humidity and temperature levels in its habitat. Preserve plant diversity within the habitat to enhance ecosystem stability.

076 中文名 **火烧兰**
学　名 *Epipactis helleborine*

兰科
Orchidaceae

火烧兰属
Epipactis

[形态描述] 地生草本，高20～70厘米；茎具2～3枚鳞片状鞘。叶4～7枚，互生；叶片卵圆形，先端通常渐尖；向上叶逐渐变窄而成披针形。总状花序，通常具3～40朵花；花苞片叶状；花梗和子房具黄褐色绒毛；花绿色或淡紫色，较小；中萼片卵状披针形，舟状；侧萼片斜卵状披针形，先端渐尖；花瓣椭圆形，先端急尖或钝；唇瓣中部明显缢缩；下唇兜状；上唇近三角形或近扁圆形，先端锐尖，近基部两侧各有一枚半圆形褶片。蒴果倒卵状椭圆状。花期7月，果期9月。

[北京分布] 房山区（上方山）；门头沟区（百花山）。

[保护策略] 保护其原生栖息地，避免过度开发；限制野生植株的采集；保持栖息地内湿度和温度相对稳定；维持栖息地内植物多样性，增强生态系统稳定性。

Morphological Description: Terrestrial herbs, 20–70 cm tall; stem with 2–3 scaly sheaths. Leaves 4–7, alternate; leaf blades ovate-elliptic, with the apex usually acuminate; upper leaves gradually narrowing to lanceolate. Inflorescence a raceme, typically with 3–40 flowers; floral bracts leaf-like; pedicels and ovary with yellow-brown pubescence; flowers small, green or pale purple; dorsal sepal ovate-lanceolate, boat-shaped; lateral sepals oblique ovate-lanceolate, acuminate at the apex; petals elliptical, with an acute or obtuse apex; lip with a distinct constriction at the center; lower lip tubular; upper lip nearly triangular or nearly flat-rounded, with an acute apex and a semi-circular lobe on each side near the base. Capsule obovate-ellipsoid. Fl. Jul, fr. Sep.

Distribution in Beijing: Fangshan District: Shangfang Mountain; Mentougou District: Baihua Mountain.

Protection Strategies: Protect its native habitat, avoid overexploitation. Restrict the collection of wild plants. Maintain relatively stable humidity and temperature levels in its habitat. Preserve plant diversity within the habitat to enhance ecosystem stability.

077

中文名 **绥草**
学　名 *Spiranthes sinensis*

兰科
Orchidaceae

绥草属
Spiranthes

[形态描述]　植株高13～30厘米。根指状。茎近基部生2～5枚叶。叶片宽线形，基部收狭，具柄状抱茎的鞘。花茎上部被腺状柔毛；总状花序具多数密生的花，呈螺旋状扭转；花苞片卵状披针形，先端长渐尖；子房纺锤形，被腺状柔毛；花小，紫红色、粉红色或白色，在花序轴上呈螺旋状排生；萼片的下部靠合，中萼片狭长圆形，舟状，与花瓣靠合呈兜状；侧萼片披针形，先端稍尖；花瓣斜菱状长圆形，先端钝；唇瓣宽长圆形，先端极钝，前半部上面具长硬毛，且边缘具强烈皱波状啮齿，唇瓣基部凹陷呈浅囊状。花期7～8月。

[北京分布]　密云区；房山区；延庆区；门头沟区。

[保护策略]　保护其原生栖息地，避免过度开发；限制野生植株的采集；保持栖息地内湿度和温度相对稳定；维持栖息地内植物多样性，增强生态系统稳定性。

Morphological Description: Plants 13–30 cm tall. The roots are finger-like. The stem has 2–5 leaves near the base. The leaf blades are broad-lanceolate, with the base narrowing into a petiole-like sheath that clasps the stem. The flower stem is covered with glandular soft hairs on the upper part. The inflorescence is a racemose cluster of numerous densely arranged flowers, spirally twisted. The floral bracts are ovate-lanceolate, with the apex gradually acuminate. The ovary is fusiform and covered with glandular soft hairs. The flowers are small, purple-red, pink, or white, arranged spirally on the inflorescence axis. The lower part of the sepals is fused; the median sepal is narrow-elliptic, boat-shaped, and fused with the petals to form a hood. The lateral sepals are lanceolate, with the apex slightly acuminate. The petals are oblique rhombic-elliptic, with a blunt apex. The lip is wide and oblong-elliptic, with a very blunt apex; the anterior half of the lip has long stiff hairs on the upper surface, and the margin has strongly wrinkled, serrated waves; the base of the lip is concave and shallowly sac-like. Fl. Jul–Aug.

Distribution in Beijing: Miyun District; Fangshan District; Yanqing District; Mentougou District.

Protection Strategies: Protect its native habitat, avoid overexploitation. Restrict the collection of wild plants. Maintain relatively stable humidity and temperature levels in its habitat. Preserve plant diversity within the habitat to enhance ecosystem stability.

078

中文名 **对叶兰**
学　名 *Neottia puberula*

兰科
Orchidaceae

鸟巢兰属
Neottia

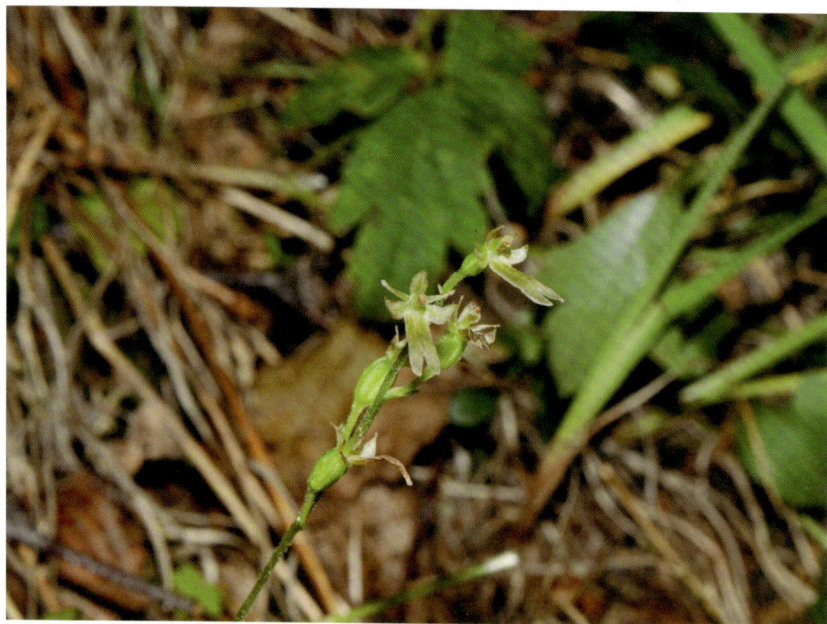

[形态描述] 植株高10~20厘米，具细长的根状茎。茎纤细，近基部处具2枚膜质鞘，近中部处具2枚对生叶，叶以上部分被短柔毛。叶片心形或宽卵形，先端急尖或钝，基部宽楔形或近心形，边缘常多少呈皱波状。总状花序被短柔毛，疏生4~7朵花；花苞片披针形，先端急尖，无毛；花梗具短柔毛；花绿色，很小；中萼片卵状披针形，先端近急尖，具1脉；侧萼片斜卵状披针形；花瓣线形，具1脉；唇瓣窄倒卵状楔形，中脉较粗，外侧边缘多少具乳突状细缘毛，先端2裂；裂片长圆形，两裂片叉开或近平行；蕊喙大，宽卵形。蒴果倒卵形。花期7~9月，果期9~10月。

[北京分布] 门头沟区（百花山）；延庆区（海坨山）。

[保护策略] 保护其原生栖息地，避免过度开发；限制野生植株的采集；保持栖息地内湿度和温度相对稳定；维持栖息地内植物多样性，增强生态系统稳定性。

Morphological Description: Plants 10–20 cm tall. Rhizome with very few elongate, filiform roots. Stem cylindric, slender, usually with 2 membranous sheaths toward base. Leaves 2, opposite, borne at ca. middle of plant, subsessile, cordate, broadly ovate, or broadly ovate-triangular, base broadly cuneate or subcordate, margin slightly crisped, apex acute or obtuse. Peduncle pubescent; rachis, pubescent, laxly 4–7-flowered; floral bracts lanceolate, glabrous, apex acute. Flowers very small, resupinate, green; pedicel pubescent; ovary pubescent; sepals and petals not spreading widely. Dorsal sepal ovate-lanceolate, 1-veined, apex subacute; lateral sepals ovate-lanceolate, oblique, apex acute. Petals linear, 1-veined, apex acute; lip narrowly obovate-cuneate or oblong-cuneate, margin slightly papillate-ciliate, apex deeply 2-lobed; lobes divergent or nearly parallel, oblong; disk with a thickened midvein. Column slightly arcuate; anther inclined toward rostellum; rostellum broadly ovate, large. Capsule obovoid. Fl. Jul–Sep, fr. Sep–Oct.

Distribution in Beijing: Mentougou District: Baihua Mountain; Yanqing District: Haituo Mountain.

Protection Strategies: Protect its native habitat, avoid overexploitation. Restrict the collection of wild plants. Maintain relatively stable humidity and temperature levels in its habitat. Preserve plant diversity within the habitat to enhance ecosystem stability.

079

<table>
<tr><td>中 文 名</td><td>**北方鸟巢兰**</td><td>兰科
Orchidaceae</td></tr>
<tr><td>学　　名</td><td>*Neottia camtschatea*</td><td>鸟巢兰属
Neottia</td></tr>
</table>

[形态描述] 植株高10~27厘米。茎上部疏被乳突状短柔毛，中部以下具2~4枚鞘，无绿叶；鞘膜质，下半部抱茎。总状花序顶生，具12~25朵花；花序轴被乳突状短柔毛；花苞片近狭卵状长圆形，膜质，背面被毛；花梗略被毛；子房椭圆形，被短柔毛；花淡绿色至绿白色；萼片舌状长圆形，先端钝，具1脉；花瓣线形，具1脉，无毛；唇瓣楔形，基部极狭，先端2深裂；裂片披针形，边缘具细缘毛；蕊柱向前弯曲；柱头凹陷，近半圆形；蕊喙大，卵状长圆形或宽长圆形。蒴果椭圆形。花果期7~8月。

[北京分布] 门头沟区（百花山）；延庆区（海坨山）。

[保护策略] 保护其原生栖息地，避免过度开发；限制野生植株的采集；保持栖息地内湿度和温度相对稳定；维持栖息地内植物多样性，增强生态系统稳定性。

Morphological Description: Plants 10–27 cm tall. The upper part of the stem sparsely covered with papilla-like short soft hairs, the lower part with 2–4 sheaths, without green leaves; sheaths membranous, the lower half clasping the stem. Terminal racemose inflorescence with 12–25 flowers; inflorescence axis covered with papilla-like short soft hairs, floral bracts lanceolate-ovate, membranous, with hairs on the back; pedicels slightly hairy; ovary elliptic, covered with short soft hairs; flowers pale green to greenish-white; sepals tongue-shaped, lanceolate, apex blunt, with one vein; petals linear, with one vein, glabrous; lip wedge-shaped, base very narrow, apex deeply bifid; lobes lanceolate, with fine marginal hairs; column curved forward; stigma concave, nearly half-circular; anther cap large, ovate-lanceolate or broad-lanceolate. Capsule elliptic. Fl. and fr. Jul–Aug.

Distribution in Beijing: Mentougou District: Baihua Mountain; Yanqing District: Haituo Mountain.

Protection Strategies: Protect its native habitat, avoid overexploitation. Restrict the collection of wild plants. Maintain relatively stable humidity and temperature levels in its habitat. Preserve plant diversity within the habitat to enhance ecosystem stability.

| 080 | 中文名 **角盘兰**
学　名 *Herminium monorchis* | 兰科
Orchidaceae

角盘兰属
Herminium |

[形态描述] 植株高5.5～35厘米。块茎球形，肉质。茎无毛，基部具2枚筒状鞘，下部具2～3枚叶，在叶之上具1～2枚苞片状小叶。叶片狭椭圆状披针形，先端急尖，基部渐狭并略抱茎。总状花序具多数花；花苞片线状披针形，子房圆柱状纺锤形，扭转，顶部明显钩曲，无毛；花小，黄绿色；中萼片椭圆形，先端钝；侧萼片长圆状披针形，先端稍尖；花瓣近菱形，中部多少3裂，中裂片线形，先端钝，具1脉；唇瓣与花瓣等长，肉质增厚，基部凹陷呈浅囊状，近中部3裂，中裂片线形，侧裂片三角形；蕊柱粗短，蕊喙矮而阔；柱头2个，位于蕊喙之下；退化雄蕊2个，近三角形。花期6～7（～8）月。

[北京分布] 门头沟区（百花山，东灵山）；延庆区（海坨山）；密云区（雾灵山）。

[保护策略] 保护其原生栖息地，避免过度开发；限制野生植株的采集；保持栖息地内湿度和温度相对稳定；维持栖息地内植物多样性，增强生态系统稳定性。

Morphological Description: Plants 5.5–35 cm tall. Tubers globose, fleshy. Stem glabrous, with 2 tubular sheaths basally, 2–3 leaves proximally, 1–2 bractlike leaflets above leaves. Leaf blade narrowly elliptic-lanceolate, apex acute, base attenuate and slightly clasping. Racemes with numerous flowers; floral bracts linear-lanceolate; ovary terete-fusiform, twisted, distinctly hooked apically, glabrous; flowers small, yellowish green; middle sepal elliptic, apex obtuse; lateral sepals oblong-lanceolate, apex slightly pointed; petals subrhombic, 3-lobed at middle, middle lobe linear, apex obtuse, with 1 vein; labellum equaling petals, fleshy and thickened, shallowly saccate at base, 3-lobed near middle, middle lobe linear, lateral lobes linear. lobes linear, middle lobe triangular, lateral lobes triangular; gynostegium stout and short; beak short and broad; stigmas 2, below beak; staminodes 2, subtriangular. Fl. Jun–Jul (–Aug).

Distribution in Beijing: Mentougou District: Baihua Mountain, Dongling Mountain; Yanqing District: Haituo Mountain; Miyun District: Wuling Mountain.

Protection Strategies: Protect its native habitat, avoid overexploitation. Restrict the collection of wild plants. Maintain relatively stable humidity and temperature levels in its habitat. Preserve plant diversity within the habitat to enhance ecosystem stability.

081

中文名 **裂瓣角盘兰**
学　名 *Herminium alaschanicum*

兰科
Orchidaceae

角盘兰属
Herminium

[形态描述]　植株高15～60厘米。块茎圆球形，肉质。茎无毛，基部具2～3枚筒状鞘，其上具2～4枚较密生的叶，在叶之上有3～5枚苞片状小叶。叶片狭椭圆状披针形，先端急尖，基部渐狭并抱茎。总状花序具多数花；花苞片披针形；子房圆柱状纺锤形，扭转，无毛；花小，绿色，中萼片卵形，先端钝，具3脉；侧萼片披针形，先端近急尖，具1脉；花瓣或多或少呈3裂，中裂片近线形，具3脉；唇瓣近长圆形，基部凹陷具距，前部3裂，侧裂片线形，中裂片线状三角形；距长圆状，向前弯曲；蕊柱粗短；蕊喙小；柱头2个，位于唇瓣基部两侧；退化雄蕊2个。花期6～9月。

[北京分布]　门头沟区（黄草梁）；延庆区（玉渡山，松山）。

[保护策略]　保护其原生栖息地，避免过度开发；限制野生植株的采集；保持栖息地内湿度和温度相对稳定；维持栖息地内植物多样性，增强生态系统稳定性。

Morphological Description: Plants 15–60 cm tall. Tubers globose, fleshy. Stem glabrous, with 2–3 tubular sheaths at the base, bearing 2–4 relatively densely arranged leaves, with 3–5 bract-like smaller leaves above the leaves. Leaves narrowly ovate-lanceolate, acute at the apex, gradually narrowing at the base and clasping the stem. Racemose inflorescence with numerous flowers; floral bracts lanceolate; ovary cylindrical, fusiform, twisted, glabrous; flowers small, green; median sepal ovate, obtuse at the apex, with 3 veins; lateral sepals lanceolate, nearly acute at the apex, with 1 vein; petals more or less 3-lobed, central lobe linear, with 3 veins; lip near oblanceolate, with a concave base and spur, trilobed at the front, lateral lobes linear, central lobe linear-triangular, spur elongated-elliptic, curving forward; column stout and short; anther cap small; stigma two, positioned on either side of the base of the lip; two reduced stamens. Fl. Jun–Sep.

Distribution in Beijing: Mentougou District: Huangcaoliang; Yanqing District: Yudu Mountain, Songshan Mountain.

Protection Strategies: Protect its native habitat, avoid overexploitation. Restrict the collection of wild plants. Maintain relatively stable humidity and temperature levels in its habitat. Preserve plant diversity within the habitat to enhance ecosystem stability.

082

中文名 **原沼兰**
学　名 *Malaxis monophyllos*

兰科
Orchidaceae

原沼兰属
Malaxis

[**形态描述**]　地生草本。假鳞茎卵形，外被白色的薄膜质鞘。叶通常1枚，卵形，先端钝或近急尖，基部收狭成柄；叶柄多少鞘状，抱茎或上部离生。花葶除花序轴外近无翅；总状花序具数十朵或更多的花；花苞片披针形，花小，淡黄绿色至淡绿色；中萼片披针形，先端长渐尖，具1脉；侧萼片线状披针形，亦具1脉；花瓣近丝状；唇瓣先端骤然收狭而成线状披针形的尾；唇盘中央略凹陷，两侧边缘肥厚并具疣状突起，基部两侧有一对钝圆的短耳；蕊柱粗短。蒴果倒卵形或倒卵状椭圆形。花果期7~8月。

[**北京分布**]　门头沟区（百花山）；延庆区（海坨山）；密云区（雾灵山）。

[**保护策略**]　保护其原生栖息地，避免过度开发；限制野生植株的采集；保持栖息地内湿度和温度相对稳定；维持栖息地内植物多样性，增强生态系统稳定性。

Morphological Description: Plants terrestrial. Pseudobulbs ovoid, relatively small, enclosed in white membranous sheaths. Leaf usually 1, ovate, oblong, or subelliptic, base contracted into amplexicaul petiole, apex obtuse or subacute. Inflorescence erect, many flowered; narrowly winged; lanceolate. Flowers pale yellowish green to pale green, small. Dorsal sepal lanceolate or narrowly ovate-lanceolate, 1-veined, apex long acuminate; lateral sepals linear-lanceolate, 1-veined. Petals filiform or narrowly lanceolate; lip ovate-triangular, caudate; disk broadly ovate or oblate, slightly concave, margin thickened and verrucose, base with a pair of short auricles on each side. Column stout. Capsule obovoid or obovoid-ellipsoid; fruiting pedicel. Fl. and fr. Jul-Aug

Distribution in Beijing: Mentougou District: Baihua Mountain; Yanqing District: Haituo Mountain; Miyun District: Wuling Mountain.

Protection Strategies: Protect its native habitat, avoid overexploitation. Restrict the collection of wild plants. Maintain relatively stable humidity and temperature levels in its habitat. Preserve plant diversity within the habitat to enhance ecosystem stability.

083

中 文 名 **羊耳蒜**
学　　名 *Liparis campylostalix*

兰科
Orchidaceae

羊耳蒜属
Liparis

[形态描述] 多年生草本；假鳞茎卵球形；茎下具2枚叶，卵形或卵状椭圆形，具鞘状叶柄；总状花序顶生，花序轴具翅；花淡黄绿色或带紫色；萼片条状披针形，花瓣丝状，唇瓣倒卵形，边缘有不明显的细齿；蒴果倒卵形。

[北京分布] 房山区（上方山）；门头沟区（百花山）。

[保护策略] 保护其原生栖息地，避免过度开发；限制野生植株的采集；保持栖息地内湿度和温度相对稳定；维持栖息地内植物多样性，增强生态系统稳定性。

Morphological Description: Perennial herb; pseudobulb ovoid-globose. Stem with two leaves at the base, ovate or ovate-elliptic, with a sheathing petiole. Terminal racemose inflorescence, with a winged inflorescence axis; flowers pale yellowish-green or purplish; sepals lanceolate, linear, petals filiform, lip inverted ovate with inconspicuous fine teeth along the margin; capsule inverted-ovoid.

Distribution in Beijing: Fangshan District: Shangfang Mountain; Mentougou District: Baihua Mountain.

Protection Strategies: Protect its native habitat, avoid overexploitation. Restrict the collection of wild plants. Maintain relatively stable humidity and temperature levels in its habitat. Preserve plant diversity within the habitat to enhance ecosystem stability.

084

中 文 名 **裂唇虎舌兰**

学 名 *Epipogium aphyllum*

兰科
Orchidaceae

虎舌兰属
Epipogium

[形态描述] 植株高10～30厘米，地下具珊瑚状的根状茎。茎直立，淡褐色，肉质，无绿叶，具数枚膜质鞘；鞘抱茎。总状花序顶生，具2～6朵花；花苞片狭卵状长圆形；花梗纤细；子房膨大；花黄色而带粉红色或淡紫色晕；萼片披针形，先端钝；花瓣与萼片相似；唇瓣近基部3裂；侧裂片直立，近长圆形或卵状长圆形；中裂片卵状椭圆形，先端急尖，边缘近全缘并多少内卷，内面常有4～6条紫红色的纵脊；距粗大；蕊柱粗短。花期8～9月。

[北京分布] 门头沟区（百花山）；延庆区（海坨山）；密云区（雾灵山）。

[保护策略] 保护其原生栖息地，避免过度开发；限制野生植株的采集；保持栖息地内湿度和温度相对稳定；维持栖息地内植物多样性，增强生态系统稳定性。

Morphological Description: Plants 10–30 cm tall, with coral-like rhizomes. The stem is erect, light brown, fleshy, without green leaves, and bears several membranous sheaths that clasp the stem. The terminal racemose inflorescence contains 2–6 flowers. Floral bracts are narrowly ovate-lanceolate. The pedicels are slender, and the ovary is enlarged. The flowers are yellow with pink or pale purple tinges. The dorsal sepal is lanceolate with a blunt apex. The petals resemble the sepals. The lip is nearly trilobed, with upright lateral lobes that are narrowly elliptic or ovate-lanceolate, and a middle lobe that is ovate-elliptic, acutely pointed, with a nearly entire margin and slightly inward-curved edges, often bearing 4–6 purple-red longitudinal ridges on the inner surface. The spur is robust, and the column is short and thick. Fl. Aug–Sep.

Distribution in Beijing: Mentougou District: Baihua Mountain; Yanqing District: Haituo Mountain; Miyun District: Wuling Mountain.

Protection Strategies: Protect its native habitat, avoid overexploitation. Restrict the collection of wild plants. Maintain relatively stable humidity and temperature levels in its habitat. Preserve plant diversity within the habitat to enhance ecosystem stability.

085

中 文 名 **北方盔花兰**

学 名 *Galearis roborowskyi*

兰科
Orchidaceae

盔花兰属
Galearis

[形态描述] 植株高5~15厘米。根状茎狭圆柱状，伸长，平展，肉质。茎直立，基部具2~3枚筒状鞘，鞘之上具叶。叶基生，卵形，先端钝或稍尖，基部收狭成抱茎的柄。花序具1~5朵花，花序轴无毛；花苞片披针形，先端渐尖；子房纺锤形，扭转，无乳突；花紫红色；中萼片卵形，舟状，先端钝，具3脉，与花瓣靠合呈兜状；侧萼片偏斜，卵状长圆形，先端钝，具3脉；花瓣卵形，先端钝或急尖，具3脉，边缘无睫毛；唇瓣宽卵形，基部具距，前部3裂；侧裂片三角形，先端稍尖，边缘波状；中裂片长圆形，先端钝，无睫毛；距圆筒状，稍向前弯曲。花期6~7月。

[北京分布] 延庆区（海坨山）。

[保护策略] 保护其原生栖息地，避免过度开发；限制野生植株的采集；保持栖息地内湿度和温度相对稳定；维持栖息地内植物多样性，增强生态系统稳定性。

216

Morphological Description: Plants 5–15 cm tall. Rhizome cylindrical, elongated spreading, and fleshy. Stem erect, with 2–3 tubular sheaths at base, and leaves above the sheaths. Leaves basal, ovate, apex obtuse or slightly acute, base gradually narrowing into a petiole clasping the stem. Inflorescence with 1–5 flowers, the rachis glabrous; floral bracts lanceolate, apex acuminate. Ovary fusiform, twisted, glabrous, without papillae. Flowers purple-red; dorsal sepal ovate, boat-shaped, apex obtuse, with 3 veins, forming a hood with the petals; lateral sepals oblique, ovate-lanceolate, apex obtuse, with 3 veins. Petals ovate, apex obtuse or acuminate, with 3 veins, margin glabrous. Lip broadly ovate, with a spur at the base, trilobed at the front, lateral lobes triangular, apex slightly acute, margin undulate, median lobe lanceolate, apex obtuse, glabrous. Spur cylindrical, slightly curved forward. Fl. Jun–Jul.

Distribution in Beijing: Yanqing District: Haituo Mountain.

Protection Strategies: Protect its native habitat, avoid overexploitation. Restrict the collection of wild plants. Maintain relatively stable humidity and temperature levels in its habitat. Preserve plant diversity within the habitat to enhance ecosystem stability.

086 中文名 **河北盔花兰**
学　名 *Galearis tschiliensis*

兰科
Orchidaceae

盔花兰属
Galearis

[形态描述] 植株高6~15厘米。根状茎指状，肉质，匍匐。茎直立，圆柱形，基部具2枚筒状鞘，鞘之上具叶。叶1枚，基生，叶片匙形，先端钝，基部收狭，具与叶片近等长的柄，抱茎。花茎无毛，花序具1~6朵花，花序轴无毛；花苞片披针形，先端渐尖；子房圆柱状纺锤形，扭转，无乳突；花紫红色、淡紫色或白色；萼片长圆形，先端钝，具3脉；中萼片凹陷呈舟状，与花瓣靠合呈兜状；花瓣直立，偏斜，长圆状披针形，边缘无睫毛，先端急尖，具2脉；唇瓣卵状披针形，基部稍凹陷，无距，先端钝或近急尖，边缘全缘或稍波状，无睫毛。花期6~8月。

[北京分布] 门头沟区（东灵山）；延庆区（海坨山）。

[保护策略] 限制开发活动，确保栖息地的稳定性，避免人为干扰；确保土壤的适度湿润；在栖息地中保留其他阴生植物，以形成良好的微气候；限制对野生河北盔花兰的采集。

Morphological Description: Plants 6–15 cm tall. Rhizomatous. Stem erect, cylindrical, with 2 tubular sheaths at the base, above which leaves are present. One basal leaf, spoon-shaped, with a blunt apex, the base gradually narrowing into a petiole of approximately equal length to the leaf blade, clasping the stem. Peduncle glabrous; inflorescence with 1–6 flowers, the rachis glabrous. Floral bracts lanceolate, tapering to a sharp point. Ovary spindle-shaped, cylindrical, twisted, glabrous. Flowers purple-red, pale purple, or white. Sepals oblong, blunt at the apex, with 3 veins; dorsal sepal concave and boat-shaped, closely appressed to the petals forming a hood. Petals erect, oblique, lanceolate-elliptic, glabrous along the edges, sharply pointed at the apex, with 2 veins. Lip ovate-lanceolate, slightly concave at the base, without a spur, with a blunt or nearly sharp apex, entire or slightly wavy at the margin, glabrous. Fl. Jun–Aug.

Distribution in Beijing: Mentougou District: Dongling Mountain; Yanqing District: Haituo Mountain.

Protection Strategies: Limit development activities to ensure habitat stability and avoid human interference. Maintain moderate soil moisture. Retain other shade-loving plants in the habitat to create a favorable microclimate. Restrict collection of wild *Galearis tschiliensis*.